Canteens and food services in industry:
A manual

Canteens and food services in industry: A manual

International Labour Office Geneva

ISBN 92-2-106637-1

First published 1988

ILO publications can be obtained through major booksellers or ILO local offices in many countries, or direct from ILO Publications, International Labour Office, CH-1211 Geneva 22, Switzerland. A catalogue or list of new publications will be sent free of charge from the above address.

Printed by the International Labour Office, Geneva, Switzerland

PREFACE

The importance of a proper, adequate and balanced diet both in the context of general health and in relation to work would hardly need emphasis. Good nutrition not only contributes to improved health and greater welfare but also provides for improved work performance and productivity. The impact can be especially significant in developing countries where workers only too often have a poor and inadequate diet.

The ILO has been concerned for well over half a century with the subject of nutrition of workers. In 1930, the first edition of the Encyclopaedia of Occupational Health and Safety contained an article on Food and Human Nutrition. In 1936 the first of many reports on "Workers' Nutrition and Social Policy" was published, and ten years later - in 1946 - the ILO published Nutrition in Industry, a book dealing with the feeding of workers in large enterprises in Canada, USA and Great Britain. The nutrition of workers has continued to be a subject of discussion in the International Labour Conference, ILO Regional Conferences and various ILO industrial and analogous committees. In 1956 the Conference adopted the Welfare Facilities Recommendation, 1956 (No. 102), which highlighted the importance of canteen and other feeding facilities for workers and specified guidelines in their establishment.

ILO's experience in this area has shown that although catering services at the plant level in industrialised countries are not yet fully satisfactory in all cases, the problem is especially serious in developing countries. Workers' catering services that do exist in developing countries are usually found in large undertakings. They are singularly lacking in smaller undertakings, although it is here that they are most needed. Another important obstacle is the lack of adequate information and expertise to advise and guide employers in setting up and operating canteens and inadequate training in industrial canteen management.

As early as 1971 a joint FAO/ILO/WHO meeting recommended, among others, the need to intensify efforts and activities to increase awareness and to prepare a simple and practical manual covering the basic principles of workers' feeding and methods for the establishment and operation of workers' feeding programmes. Hence this manual which we hope will meet this long felt need and promote greater interest and effort by national authorities, employers and workers in the development, management and operation of canteens and other types of food services for workers.

C. Dumont,
Chief,
Conditions of Work and Welfare
Facilities Branch,
Working Conditions and Environment
Department.

Table of contents

Page

INTRODUCTION 1

PART I: BACKGROUND 3

Chapter 1: Standards and government requirements 5

Chapter 2: Nutrition 11

PART II: TYPES OF FOOD SERVICES 19

Chapter 3: Mess rooms 21

Chapter 4: Refreshment facilities and mobile food vans 25

Chapter 5: Local vendors 27

Chapter 6: Canteens including inter-enterprise canteens 29

Chapter 7: Low-cost shops 34

PART III: ORGANISATION, MANAGEMENT AND FINANCING OF CANTEENS 39

Chapter 8: Organisation and administration 41

Chapter 9: Methods of financing 56

Chapter 10: Physical facilities 66

Chapter 11: Equipment and maintenance 77

Chapter 12: Operating the canteen 86

CONCLUSION 95

ANNEXES 99

Annex I : Check list 101

Annex II : Cook/freeze and cook/chill methods of cooking 111

Annex III : Bibliography 113

INTRODUCTION

This manual has been designed to assist government officials, labour inspectors, employers, personnel managers and workers' representatives in developing countries to determine what type of food service would best serve the needs of workers in various circumstances. It provides suggestions on how to upgrade or expand existing services and on how to set up new ones.

Although it is not intended to be a text-book of industrial catering, the manual may serve as a reference guide for catering managers and may assist in training new canteen staff.

In many countries, enterprises are required by law to set up mess rooms (known as lunch or crib rooms in some countries) or canteens depending on the size of the undertaking. However, employers often do not have the expertise on their staff to set up food services and find it difficult to locate personnel with relevant experience in catering or similar fields to advise them. This manual gives practical advice on the establishment of a variety of food services. It progresses from a discussion of low-cost services for small and medium-sized enterprises - such as mess rooms, mobile food vans or food stalls set up near the enterprise by vendors - to the administration of canteens.

The manual discusses factors that should be taken into account in determining what service to provide, such as the number of workers; any legal requirements to provide a canteen or mess room; hours of work, local eating habits, whether and by how much the meals will be subsidised; whether the workers prefer to bring their own meals from home and the availability of other food from vendors, food stalls and restaurants near the workplace.

Structure of the manual

A brief discussion of standards and government regulations and nutrition is provided in Part I. Part II considers the operation of mess rooms, refreshment facilities and mobile vans, the provision of facilities for local vendors, canteens including inter-enterprise canteens, and low-cost shops. Part III focuses on canteens. It provides information on their organisation and administration, financing, physical facilities, equipment and maintenance and management. Such aspects as the planning of food services, training of staff, financial control, physical facilities and making use of available skills and talents within the enterprise are covered.

The emphasis throughout is on providing the best food service possible using the personnel and the facilities available or easily obtainable locally while taking into consideration the nutritional needs of workers. It is hoped that the information provided will assist in the operation of successful food services and canteens which provide adequate, healthy nutrition for all those who use them.

PART I: BACKGROUND

Chapter 1

Standards and government requirements

Many countries have legal requirements for the provision of canteens, mess rooms (known as lunch room or crib room in some countries) or other food services. Some national laws stipulate that there must be a mess room where more than 100 workers are employed, for example, and a canteen with cafeteria where more than 250 workers are employed. Legislation in some countries requires the provision of mess rooms for workers employed in dirty, dangerous or toxic jobs.

The setting up of canteens in developing countries is not always the result of legislative requirements or purely voluntary action by the employer. In a few countries, workers have made the provision of a canteen one of their claims in collective bargaining. Occasionally, collective agreements contain clauses concerning conditions to be observed in existing canteens, including workers' participation in the management of the canteen, the level of prices to be charged, and the quantity and quality of food to be served.

Several countries have standards concerning occupational safety and health which prescribe measures in respect of premises, equipment and cleanliness in canteens and kitchens and may also require the employment of a trained dietitian. Some governments have policies on the question of nutrition in relation to canteens.

Regulations may be detailed, specifying how often the kitchen walls are to be lime-washed or washed down with soap and water; the type of finish of the lower part of the kitchen walls; and the type of table-tops to be used in the cafeteria. Such regulations refer to normal hygienic practices which would be carried out in most enterprises today without the need for regulations. In the past, however, these standards were less widely understood and practised and so it was necessary to write them into the law.

Government regulations on hours of work, holidays, etc., should be observed in the canteen unless there is a specific exemption. One result of this is that more regular working hours in industrial catering may appeal to older workers who no longer find that the so-called glamour of working in hotels for higher wages compensates for the longer hours and lack of homelife which is so often found in hotel work.

The keeping of official records on such matters as hours of work, holidays, accidents and dangerous occurrences is as necessary for a canteen as for any other workplace. However, since a small canteen may only be able to employ a cook-supervisor and not a manager, it may well be necessary to provide this supervisor with clerical help.

Hygiene, safety and health requirements

Occupational safety and health regulations apply in the canteen and mess room. The safety professional, if one is employed, should make inspections and give advice and training on safe working to the kitchen and cafeteria staff. His or her advice should be freely sought since hazards, such as improperly guarded items of equipment or wet, slippery floors, can occur in the kitchen as well as elsewhere.

Additionally, and equally important, there is the requirement to conform to the regulations dealing with hygienic handling of food. The medical officer, or an industrial nurse if there is one, should make inspections and be available to give advice whenever a health problem arises.

It is advisable to invite the government health inspector or environmental health officer to visit the canteen before it opens, so that he or she can make suggestions at that stage rather than after the canteen is in operation. Subsequent visits may be frequent or infrequent depending upon the inspector's work arrangements. In a well-run kitchen the supervisor or manager should be able to spot infringements of the safety and food hygiene regulations and take immediate corrective action.

The safety and food hygiene regulations should be posted up on the wall of the canteen. If the language of the regulations is not readily understood by all the kitchen staff, a translation into a language they understand should be added alongside the original.

National and international standards

Many countries have national institutes for standards which lay down technical standards for kitchens and equipment. The International Organization for Standardization (ISO) in Geneva, Switzerland, co-ordinates the work of the national bodies and also draws up some international standards. It is advisable to verify that all equipment purchased conforms to the national standard if there is one and when importing equipment to check that it conforms to the current standards of the exporting country. There have been isolated instances of equipment which no longer met the safety requirements of the manufacturing country being sold to unsuspecting purchasers in a developing country.

Standards for purity and quality of foodstuffs

The purity and quality of the food purchased by the canteen should normally be assured by governmental controls, which are very strict in some countries. There are still, unfortunately, some countries where adulteration of food at the retail level is said to be common. The risk of purchasing such adulterated food can be minimised by buying directly from wholesalers. However, this may not always be convenient or even possible for some small canteens since their storage space is so small that their orders are too small for a wholesaler.

Contamination of foodstuffs by pesticides should also be guarded against. If there are no government standards on this risk, the canteen manager or supervisor should make enquiries about the farming practices of suppliers of fruit and vegetables, and should let it be known that the canteen is not interested in purchasing food contaminated by pesticides. Visits to the canteen's food suppliers are worthwhile as they demonstrate interest and result in the canteen receiving better service.

ILO Welfare Facilities Recommendation, 1956 (No. 102)

When new legislation is drafted on food services or existing legislation is revised or amended, the section on feeding facilities of the Welfare Facilities Recommendation might be used as an example to follow. This section is reproduced below.

RECOMMENDATION CONCERNING WELFARE FACILITIES FOR WORKERS

The General Conference of the International Labour Organisation,

Having been convened at Geneva by the Governing Body of the International Labour Office, and having met in its Thirty-ninth Session on 6 June 1956, and

Having decided upon the adoption of certain proposals with regard to welfare facilities for workers, which is the fifth item on the agenda of the session, and

Having determined that these proposals shall take the form of a Recommendation,

adopts this twenty-sixth day of June of the year one thousand nine hundred and fifty-six the following Recommendation, which may be cited as the Welfare Facilities Recommendation, 1956:

Whereas it is desirable to define certain principles and establish certain standards concerning the following welfare facilities for workers:

(a) feeding facilities in or near the undertaking;

(b) rest facilities in or near the undertaking and recreation facilities excluding holiday facilities; and

(c) transportation facilities to and from work where ordinary public transport is inadequate or impracticable.

The Conference recommends that the following provisions should be applied as fully and as rapidly as national conditions allow, by voluntary, governmental or other appropriate action, and that each Member should report to the International Labour Office as requested by the Governing Body concerning the measures taken to give effect thereto.

I. Scope

1. This Recommendation applies to manual and non-manual workers employed in public or private undertakings, excluding workers in agriculture and sea transport.

2. In any case in which it is doubtful whether an undertaking is one to which this Recommendation applies, the question should be settled either by the competent authority after consultation with the organisations of employers and workers concerned, or in accordance with the law or practice of the country.

(continued)

II. Methods of implementation

3. Having regard to the variety of welfare facilities and of national practices in making provision for them, the facilities specified in this Recommendation may be provided by means of public or voluntary action -

(a) through laws and regulations, or

(b) in any other manner approved by the competent authority after consultation with employers' and workers' organisations, or

(c) by virtue of collective agreements or as otherwise agreed upon by the employers and workers concerned.

III. Feeding facilities

A. Canteens

4. Canteens providing appropriate meals should be set up and operated in or near undertakings where this is desirable, having regard to the number of workers employed by the undertaking, the demand for and prospective use of the facilities, the non-availability of other appropriate facilities for obtaining meals and any other relevant conditions and circumstances.

5. If canteens are provided by virtue of national laws or regulations, the competent authority should be empowered to require the setting up and operation of canteens in or near undertakings where more than a specified minimum number of workers is employed or where this is desirable for any other reason determined by the competent authority.

6. If canteens are the responsibility of works committees established by national laws or regulations, this responsibility should be exercised in undertakings where the setting up and operation of such canteens are desirable.

7. If canteens are provided by virtue of collective agreements or in any other manner except as indicated in Paragraphs 5 and 6, the arrangements so arrived at should apply to undertakings where this is desirable for any reason as determined by agreement between the employers and workers concerned.

8. The competent authority or some other appropriate body should make suitable arrangements to give information, advice and guidance to individual undertakings with respect to technical questions involved in the setting up and operation of canteens.

9. (1) Where adequate publications are not already in existence, the competent authority or some other appropriate body should prepare and publish detailed information, suggestions and guidance, adapted to the special conditions in the country concerned, on methods of setting up and operating canteens.

(continued)

(2) Such information should include suggestions on -

(a) location of the canteens in relation to the various buildings or departments of the undertakings concerned;

(b) establishment of joint canteens for several undertakings in so far as is appropriate;

(c) accommodation in canteens: standards of space, lighting, heating, temperature and ventilation;

(d) layout of canteens: dining room or rooms, service area, kitchen, dishwashing area, storage, administration office, and lockers and washroom for canteen personnel;

(e) equipment, furnishing and decoration of canteens: equipment for the preparation and cooking of food, refrigeration, storage and washing up; types of fuel for cooking; types of tables and chairs in the dining room or rooms; scheme of painting and decoration;

(f) types of meals provided: standard menu, standard menu with options, _à la carte_; dietetic menus where medically prescribed; special menus for workers in unhealthy occupations; breakfast, midday meal or other meals for shift workers;

(g) standards of nutrition: nutritional values of foodstuffs, planned menus and balanced diets;

(h) types of service in the canteen: hatch or counter service, cafeteria, and table service; personnel needed for each type of service;

(i) standards of hygiene in the kitchen and dining rooms;

(j) financial questions: initial capital outlay for construction, equipment and furnishing, continuing overheads and maintenance expenses, food and personnel costs, accounts, prices charged for meals.

B. Buffets and trolleys

10. (1) In undertakings where it is not practicable to set up canteens providing appropriate meals and in other undertakings where such canteens already exist, buffets or trolleys should be provided, where necessary and practicable, for the sale to the workers of packed meals or snacks and tea, coffee, milk and other beverages. Trolleys should not, however, be introduced into workplaces in which dangerous or harmful processes make it undesirable that workers should partake of food and drink there.

(2) Some of these facilities should be made available not only during the midday or midshift interval but also during the recognised rest pauses and breaks.

(continued)

C. Messrooms and other suitable rooms

11. (1) In undertakings where it is not practicable to set up canteens providing appropriate meals, and, where necessary, in other undertakings where such canteens already exist, messroom facilities should be provided, where practicable and appropriate, for individual workers to prepare or heat and take meals provided by themselves.

(2) The facilities so provided should include at least -

(a) a room in which provision suited to the climate is made for relieving discomfort from cold or heat;

(b) adequate ventilation and lighting;

(c) suitable tables and seating facilities in sufficient numbers;

(d) appropriate appliances for heating food and beverages;

(e) an adequate supply of wholesome drinking water.

D. Mobile canteens

12. In undertakings in which workers are dispersed over wide work areas, it is desirable, where practicable and necessary, and where other satisfactory facilities are not available, to provide mobile canteens for the sale of appropriate meals to the worker.

E. Other facilities

13. Special consideration should be given to providing shift workers with facilities for obtaining adequate meals and beverages at appropriate times.

14. In localities where there are insufficient facilities for purchasing appropriate food, beverages and meals, measures should be taken to provide workers with such facilities.

F. Use of facilities

15. The workers should in no case be compelled, except as required by national laws and regulations for reasons of health, to use any of the feeding facilities provided.

Welfare Facilities Recommendation, 1956 (No. 102)

Chapter 2

Nutrition

Energy requirements of men and women

When food is freely available, workers in good health will normally eat enough to meet or even exceed their individual requirements for energy. If they consistently eat more or less than they need for their life processes and for the physical activities of their work and recreational activities, their body weight and body composition will change. If they use up more energy than their food intake provides, their reserves of body fat and other tissues will be used to make up the difference, with the result that there is a loss of weight and a lessened ability to maintain sustained physical effort. Undernourished men and women tend to be less resistant to disease and infection than the well nourished. On the other hand, if the food they eat consistently supplies more energy than they require, the unused energy accumulates as body fat.

As this is not a textbook on nutrition and dietetics, much detail on energy requirements has deliberately been omitted. Readers will find such information in textbooks on nutrition and dietetics, some of which are listed in the bibliography. Nutritionists and physiologists use the terms kilocalories (kcal) and kiloJoules (kJ) to measure both the energy expended in a particular activity and the energy provided by a certain amount of a particular food. (There are approximately 4.2 kilocalories in a kiloJoule but the two terms are not truly comparable in scientific terms.) In this manual the terms "calorie" and "energy value" have been used in the text as these may be more familiar terms to many readers.

However, Table A below shows the energy expenditure for certain activities in both kcal and kJ per minute. This and other tables should be used with caution, since the work performed may vary considerably from case to case. An extreme example is the difference in energy expenditure between work in old-fashioned steel rolling mills and work in new mills with computer-controlled equipment and air-conditioned control rooms.

Protein supplements

The protein content of the daily diet is important - see below under "Body-building foods". There is little need to give protein supplements to workers doing heavy manual work, provided their energy intake is adequate. It may, however, be helpful or even necessary to supply extra protein and other nutrients, as well as extra energy food in order to restore the muscle mass and energy reserves of workers who have experienced long periods of illness or undernutrition.

Classification of foods

Foods are normally divided into the following groups: energy-giving foods; body-building and maintenance foods (protein); and protective foods.

TABLE A - ENERGY EXPENDITURE IN SPECIFIED ACTIVITIES - MEN

	kcal per min.	kJ per min.
In bed asleep or resting	1.08	4.52
Sitting quietly	1.39	5.82
Standing quietly	1.75	7.32
Walking 3 miles/hour (4.9 km/hour)	3.7	15.5
Walking 3 miles/hour (4.9 km/hour) with a 10-kg load	4.0	16.7
Office work (sedentary)	1.8	7.5
Domestic work		
Cooking	2.1	8.8
Light cleaning	3.1	13.0
Moderate cleaning (polishing, window cleaning, chopping firewood, etc.)	4.3	18.0
Light industry		
Printing	2.3	9.6
Tailoring	2.9	12.1
Shoemaking	3.0	12.6
Garage work (repairs)	4.1	17.2
Carpentry	4.0	16.7
Electrical industry	3.6	15.1
Machine tool industry	3.6	15.1
Chemical industry	4.0	16.7
Laboratory work	2.3	9.6
Transport		
Driving lorry	1.6	6.7
Building industry		
Labouring	6.0	25.1
Bricklaying	3.8	15.9
Joinery	3.7	15.5
Decorating	3.2	13.4
Farming (tropical)		
Grass cutting (with cutlass)	4.5	18.8
Clearing bush	6.2	25.9
Planting	3.6	15.1
Weeding (Africa)	3.8-7.8	15.9-32.6
Ridging, deep digging (Africa)	5.5-15.2	23.0-63.6
Tree felling (Africa)	8.4	35.1
Head planning, 20-35 kg loads (Africa)	3.2-5.6	13.4-23.4
Mowing (India)	5.1-7.9	21.3-33.0
Watering (India)	4.1-7.5	17.1-31.4
Weeding, digging, and transplanting	2.3-9.1	9.6-38.1
Farming		
Driving tractor	2.4	10.0
Forking	7.8	32.6
Loading sacks	5.4	22.6
Feeding animals	4.1	17.2
Repairing fences	5.7	23.8

TABLE A (Cont'd)

	kcal per min.	kJ per min.
Forestry		
In nursery	4.1	17.2
Planting	4.7	19.7
Felling with axe	8.6	36.0
Trimming	8.4	35.1
Sawing - hand saw	8.6	36.0
power saw	4.8	20.1
Mining		
Working with pick	6.9	28.9
Shovelling	6.5	27.2
Erecting roof supports	5.6	23.4
Recreations		
Sedentary	2.5	10.5
Light (billiards, bowls, cricket, golf, sailing, etc.)	2.5-5.0	10.5-21.0
Moderate (canoeing, dancing, horse-riding, swimming, tennis, etc.)	5.0-7.5	21.0-31.5
Heavy (athletics, football, rowing)	7.5 +	31.5 +

Source: FAO - Food and Nutrition Paper No. 6: "The feeding of workers in developing countries", Rome, FAO, 1986.

Energy-giving foods include carbohydrates, fats and oils. Carbohydrate foods include: sugar, rice, wheat, bread, sago, semolina, oatmeal, barley, cassava, pasta, root vegetables, dried fruits such as dates and figs, wheat flour, cornflour, pulses, dhals, beans, peas, lentils, maize and fruit, in particular bananas. Fats and oils include butter, ghee, margarine, red palm oil, olive oil, sunflower oil, lard, fatty meat, fatty fishes, cod liver oil, mustard oil and nuts.

Body-building foods are essential to the building of bones and muscles etc., and are also essential in adult life to maintain the body in good condition.

Young people, pregnant women and nursing mothers, and persons recovering from long illness or undernutrition have increased protein requirements. Protein foods can be divided into animal protein and vegetable protein. The animal proteins contain all the amino acids necessary for growth and maintenance and so are sometimes called complete proteins, while the vegetable proteins do not contain all necessary amino acids and are sometimes called incomplete proteins. However, a mixture of vegetable proteins can provide a complete protein, since the amino acids missing in one vegetable may be found in another and there is a complementary effect. Long before scientists had begun to unravel the complexities of amino acids and proteins, mankind had discovered the benefits of mixing vegetable proteins, such as dhal and rice.

There is also a complementary effect if a small amount of animal protein is taken at the same meal as a larger amount of vegetable protein. Animal protein sources include: fish, all kinds of meat, poultry and game, shellfish, eggs, milk, cheese and curds. Vegetable protein sources include: pulses, semolina, soya, nuts, rice, oatmeal, wheat and products made from wheat, and to a lesser extent root and leafy vegetables. Vegetables provide a significant amount of protein in a vegetarian diet because of the quantities eaten.

Protective foods are foods that contain the vitamins and mineral salts necessary for complete health. All vegetables, fruits and animal proteins and many fats and oils come into this group, and good diet contains a mixture of them.

Textbooks on dietetics and nutrition deal with this subject in great depth but from the practical angle of a small enterprise providing an industrial feeding service there are some simple rules which help to ensure that the food provided contains nutrients from the protective and body-building groups:

- Does the meal on the plate have natural colour? (Yellow and green vegetables and yellow and red fruit contain vitamins A and C and some of the B-complex vitamins).

- Does the meal on the plate contain a mixture of foods, such as dhal, chapatti, curried vegetables?

- Are small portions of expensive animal protein foods supplemented by vegetable proteins?

- Is the frying oil used of good quality? (It is important not to re-use oil too many times).

The canteen staff should cook green vegetables as briefly as possible in small quantities of water and use the cooking water in stews, curries, etc. in order to retain as many vitamins as possible.

The supply of safe drinking water - which is clean and uncontaminated - is vital, especially in hot climates, underground mines and near furnaces. Whether salt tablets or saline solutions need to be taken depends on many factors, including the type of diet normally eaten. The medical officer's advice should be sought.

The prevention and diagnosis of illnesses caused by food contaminants (including bacteria, mycotoxins, chemicals and heavy metals) requires skilled staff and specially equipped laboratories and is the work of Public Health Departments who will advise on specific problems. At enterprise level the risk of contracting food-borne diseases (which cause suffering, lost working time and reduced work capacity) can be minimised -

(a) by medical supervision of all food handlers, and by prompt and efficient treatment when they are ill;

(b) by strict cleanliness in all food stores, kitchens, messrooms, cafeterias, hawkers' stalls, kiosks and cafes;

(c) by supplying safe, pure and wholesome drinking water. If there is no piped drinking water available in the enterprise, the firm should ensure that all drinking water, whether from a company well or a water-carrier, is regularly analysed and controlled in an approved laboratory. If necessary the drinking water should be filtered and/or chemically treated or boiled on company premises. The advice of the local public health authorities should be obtained;

(d) the sources of milk, meat, poultry, fish and shellfish should be checked by responsible person(s) before a purchasing contract is signed and during the course of a contract.

Special care is needed when buying meat in rural areas since unhealthy animals are sometimes slaughtered and their meat sold for human consumption. Such meat has caused illness in those who ate it. Contaminated and adulterated milk cause ill-health and in general it is advisable to boil it before use. Stringent precautions are needed (sealed cans, use of hygrometers, etc.) to prevent adulteration. Fish and shell-fish should come from clean waters and be handled hygienically. Smoking (fuming) and salting do not make contaminated food safe.

The role of an industrial feeding service in improving the nutrition of workers varies from country to country, but there is no doubt that a forward-looking approach to nutrition in a canteen can do much to support government-sponsored programmes, by following government recommendations and displaying posters on nutrition and ensuring that the canteen follows the principles advocated in the poster.

Where a government has published recommended dietary allowances in a simple graph showing the amount of the various main nutrients present in the food served, this could be used to provide basic information to workers and the food service staff.

Vitamin sources

A short list of good sources of the more important vitamins needed in the daily diet is given below to help those responsible for running a small canteen or industrial feeding service. Usually, it is the amount of food eaten regularly which decides whether the food is a good source of vitamin in a particular area. For example, pawpaw (papayas) are a good source of vitamins A and C in tropical countries where they are cheap and plentiful. In colder countries potatoes and winter greens are often eaten regularly and are thus a good source of vitamin C and fatty fishes are a good source of vitamin A. (The outer dark green leaves of winter greens are not often eaten in cold climates so are less likely to be a good source of vitamin A).

Vitamin A is found mainly in:

- Animal sources (as retinol)

  Whole milk (fresh or dried), butter and vitaminised margarine, ghee made from butter, cheese, yoghurt, egg yolk, some fatty fish (tuna, herring, sardine, etc.) and some fish liver oils.

- Fruit and vegetable sources (as carotene)

  Red palm oil, hominy/maize grits, olives in brine, red chilis, tomatoes, carrots, gourds, pumpkins, peas, sweet potatoes (especially red and yellow), and green leafy vegetables. Apricots, avocados, bananas, melons with yellow flesh, papayas/pawpaws, figs, guavas, ripe yellow/orange mangoes, oranges and mandarins.

Vitamin A is destroyed by cooking or soaking in oil but not by boiling, simmering or soaking in water.

Vitamin D

In places where there is plenty of sun and people do not wear the kind of clothing which prevents the sun from reaching the skin, the body is normally able to make Vitamin D from the action of the ultra-violet rays of sunlight on the skin.

Good dietary sources of Vitamin D include fatty fish, egg yolk, butter, vitaminised margarine.

Vitamins of the B complex

These include thiamin, riboflavin, pantothenic acid, nicotinic acid, folic acid and B12. They are found chiefly in -

- Green leafy vegetables
  brown and whole-grain cereals (e.g. wheat, rice, maize, etc.), pulses, (e.g. lentils, dhal, gram, chick-peas, etc.), ground-nuts, yeast.

- Animal sources

  eggs, milk, fish, meat.

Vitamin C (ascorbic acid)

all leafy green vegetables, cauliflowers, horseradish, mint, cresses and mustard leaves, peas, potatoes, and all sprouting grains or seeds such as bean sprouts.

- Fruits

  currants, black and red cranberries, gooseberries, grapefruit, oranges, lemons, limes, pomeloes, mandarins, guava, kiwi fruits, papayas/pawpaws, lychees, cumquats, longan, mangoes, melons, pineapples, quinces, raspberries, strawberries. Eat fruit raw if possible.

Since vitamin C is destroyed rapidly by heat, alkaline solutions, cooking in water, soaking in water and exposure to air it is necessary to ensure that vegetables are cut or sliced immediately before cooking, to use as little cooking liquid as possible, to cook as quickly as possible and never use bicarbonate of soda, serve quickly once cooked and to add any left-over cooking liquid to sauces, gravies or stews.

Women workers - special nutritional needs

Women workers in developing countries may suffer from undernutrition to a slight but significant degree. This occurs especially in those cultures where traditionally the women stay at home to rear the children, clean the house and tend the family farm or garden. When, however, a women goes to work in addition to her household duties, her family frequently do not understand that her energy expenditure is considerably greater than if she were to remain at home all day. She therefore needs more food if she is to be able to continue to work and remain in good health. It has been estimated that the average woman worker in the lowest socio-economic groups may require 300 more calories a day than she normally consumes.

Where an enterprise employs a majority of women workers the employer may consider providing a special snack of locally acceptable foods, containing a mixture of vegetable protein, some carbohydrate and some green leafy vegetables or fruit, even when a full industrial feeding service is not envisaged. Special attention to the nutritional needs of pregnant and nursing women should be considered. For example, milk could be provided with the snack.

The following are suggested high nutritional snacks:

> "250 grams of cooked rice, yam, potato or cassava with a fish, bean or meat sauce, wrapped in a banana leaf; 125 grams bread or roll or tortilla with an egg, cheese, meat or fish filling; 100 grams of groundnuts with a beverage.
>
> Each snack would provide roughly 400 calories.
>
> Sweetened tea or coffee or a soft drink with biscuits, although popular, cannot be considered as a substantial snack, but is nevertheless very welcome as a mid-shift refreshment break and is a good source of liquid in hot countries."*

Diet and cultural questions

It is difficult to separate diet from cultural questions as it is important to consider cultural needs and requirements when groups of people of differing backgrounds, religious beliefs and ages have to eat together.

These needs and requirements should be considered when planning the feeding service, designing the cafeteria, planning menus and selecting staff to work in the canteen.

There are however some simple questions to be asked before a canteen is opened, and again at intervals when it is in service. The canteen planning committee and, later, the canteen management committee should consider the answers to the following questions:

- How many religions are represented among the workers?
- Are some religious groups obliged to fast every year - for one day? for one month? If so, is it possible to make arrangements for those workers to take their meal-break period away from the sight of food during a fasting period?
- Are adequately varied menus provided for vegetarians?
- Are labels placed on pork and other meat dishes describing their contents?

(These questions should be asked again after the canteen has been running for a few years - the answers may be quite different).

There is, too, the desirability of supplying, if not every day, at least once a week a dish which is cooked and seasoned to the taste of minority groups.

---

* FAO-Food and Nutrition Paper No. 6. "The feeding of workers in developing countries", Rome, FAO, 1986.

Festivals and celebrations

It may well be possible for an establishment to make the cafeteria available to their workers for special festivals and celebrations. In addition, the cafeteria might be decorated appropriately for the period of the festival and one traditional festive dish offered on the menu.

Weddings and some religious ceremonies are occasions when members of a community come together and eat together. In some countries it is customary for an employer, in one way or another, to subsidise such gatherings. The subsidy may vary from a cash gift to the loan of the company cafeteria or messroom, or the provision of furniture such as tables and chairs for the ceremony.

Due consideration of the cultural backgrounds of the workers requires thought and understanding. It is comparatively easy to ensure that food preferences are met, religious dietary laws observed and the customs surrounding certain ceremonies sympathetically observed.

Summary

Hygienic practices, the elimination of intestinal worm infestations, the provision of clean drinking water and well-cooked food served in pleasant surroundings, the observance of the dietary laws of different religious and ethnic groups, and the prompt investigation of complaints about food - all these contribute to the good nutrition of the individual and must be considered when planning for the improved nutrition of workers. Special consideration should be given to the following; the nutritional requirements of workers engaged in heavy manual work and those working in extremes of hot or cold climatic conditions; the need for salt tablets or salty drinks in very hot climates or hot working environments such as in steel mills or foundries; and the need for extra liquids in hot climates. Industrial medical officers can obtain advice on assessing the nutritional requirements of workers from nutritionists and dietitians. Where an extra allowance of a particular food or beverage has traditionally been given to workers in certain jobs, it should only be withdrawn when there is clear evidence from experts that it is not beneficial to the worker and after the situation has been frankly discussed with the workers' representatives.

PART II: TYPES OF FOOD SERVICES

Chapter 3

Mess rooms

Introduction

Mess rooms (known as lunch rooms or crib rooms in some countries) are places where workers can eat and, if necessary, reheat food they have brought with them. Facilities for heating food, boiling water and preparing hot drinks, disposing of waste food, washing dishes and hand-washing should be provided, and an attendant is usually employed to keep the messroom tidy and clean.

The provision of mess rooms can play an important role in feeding the large number of workers who work in enterprises without canteens or who prefer home-cooked food.

Legal requirements

Many countries have statutory requirements for the provision of mess rooms where, for example, more than 100 workers are employed, or according to the distance of the workplace from the nearest town or village. Some countries require the provision of canteens or mess rooms for workers engaged in dangerous, dirty or toxic processes.

Location and physical facilities

In a small establishment the choice of site for a mess room will probably be limited. However since the cooking done in it will be restricted to reheating food and boiling water, the requirements for ventilation and access for garbage collectors are not so stringent as in the case of a canteen with a full kitchen and cafeteria.

A growing establishment that is planning over the next few years to build a canteen might build a mess room on the future site of the canteen and install the basic utilities at this early stage. It would then be possible to convert the mess room into a canteen when desired with a minimum of alterations and costs.

The following should be considered when building a mess room:

- Entrance door - should have a porch, double door or veranda to give protection from winds and heat. It should have a screen door if flies are numerous at any time of year.

- Veranda - useful in hot climates and also a source of extra space.

- Sanitary facilities - should be located nearby, providing toilets and wash basins so that workers can wash their hands before eating.

- Walkways - if the mess room building is separate from the rest of the establishment a covered walkway to it from the workshops and offices is desirable.

- Cloakroom - some form of secure hanging space for umbrellas and wet outdoor clothing may be needed if there are no covered walkways between the workplace and the mess room.

- Water supply - there must be an adequate supply of drinking water as well as hot and cold water sinks for washing up crockery and cutlery.

- Fuel supply - there should be a supply of fuel for cooking - gas, electricity and kerosene (paraffin) are the cleanest. If wood, coal or charcoal are used there must be a proper chimney flue (ventilation is of course important when charcoal is used indoors - charcoal should be avoided as far as possible in a mess room or for industrial catering).

- If solid fuel is used, the attendant will have to maintain the fire and make sure it is at the correct temperature at meal and break periods.

- Garbage - there should be a separate area for the garbage collection, which is protected from animals and flies and is easily cleaned. It is wise to make this large enough to hold the garbage for two weeks, even if there is normally a weekly collection, to provide for interruptions of the service or an unexpected increase in the number of workers fed, for example.

- Safety and fire prevention - the arrangement of equipment, the choice of non-slip flooring, the standard of housekeeping in the mess room - all these help reduce the level of potential hazards. As a minimum there should be a bucket of clean sand and a fire blanket at hand for fire-fighting.

Whatever the underlying reason for setting up a mess room, it inevitably involves expense for a firm, especially for a small enterprise, and money will be wasted if the workers do not make full use of it.

It is therefore useful first to determine the number of workers likely to use the mess room by carrying out a survey in co-operation with workers' representatives. The maximum number of workers that will eat at the same time will also have to be calculated to provide for an adequate number of seats and tables. The mess room can then be planned so that it becomes a place not only for eating but also for relaxation.

Furniture and equipment

One way of making the mess room acceptable to all the groups of workers in a small establishment is to stagger the meal breaks so that different groups eat at different times. Another method is to divide up the room into areas. One way of doing this is to place the equipment, water boiler, boiling rings or hot plate, sink and refrigerator in the middle of the room to provide a divider. Wash basins could be provided near the entrance door at each end. One end of the room might be used for various other activities, such as card games or table tennis. A notice-board might be placed on the wall to post notices about enterprise and local activities.

In addition to the basic requirements of tables and chairs - or clean floor mats in regions where it is customary to sit on the floor to eat - there should be a water boiler, a hot plate or small oven, a washing-up sink, at least one wash basin and a refrigerator to provide safe storage for the food brought by workers until it is eaten.

Size of mess room

Where the law does not prescribe a minimum size for a mess room the following rough guidelines for floor space may be used. If the room is irregularly shaped or has a low ceiling it is advisable to allow extra space. Staggering of meal breaks may be of help if space is limited or when it is desired that office staff should eat separately from the shop-floor workers.

Floor area for mess rooms - seating space and equipment

Up to 25 customers at one time:	14 sq.ft. per person (1.30 m^2)
25-50 customers at one time:	13 sq.ft. per person (1.20 m^2)
50-100 customers at one time:	12 1/2 sq.ft. per person (1.16 m^2).*

A mess room can be designed for a variety of purposes. It can be used for meetings and staff training, family-planning meetings, celebrations, apprentice graduation ceremonies, union meetings, films, sports training, etc. However, it should never be used as a workplace, and none of the activities mentioned above should take place at the same time as a meal break.

Consideration should be given to the use of stackable chairs and tables when planning a multi-purpose mess room.

Mess room attendant or steward

A key factor in the success of a mess room is the attendant or steward. This person is responsible for cleaning the room and maintaining it in good condition. Whoever is appointed should be conscientious and pleasant. The attendant may also take a tea trolley to the workshops and offices in the middle of the afternoon.

Mess room committees

In a small establishment a formal committee to set up and run a mess room may not be necessary. Where needed, a committee can be set up involving the representatives of management and workers.

Cleaning

Daily and weekly cleaning should be carried out by the steward or attendant. A member of the mess room management committee or senior member of staff should be responsible for checking that cleaning is done properly and for ensuring that there are adequate supplies of cleaning materials available. The walls and ceiling should be cleaned regularly, and at least once a year.

Inter-enterprise mess rooms

Where a mess room is to be used by the workers of several small enterprises, agreement must be reached as to who will be responsible for its maintenance. Perhaps an individual enterprise may offer to provide the attendant

---

* Canteens, messrooms and refreshments services. Health and Safety at Work Series No. 2, Dept. of Employment and Productivity. (HMSO, London, 1970).

and some general supervision, or a canteen management committee will undertake the day-to-day running of the mess room. Whatever arrangements are made, it will be necessary to settle financial matters before the mess room is opened.

Agreement should be reached on the opening hours. The mess room should be big enough to accommodate the largest number of persons likely to use it at any one time. In order to avoid overcrowding, an agreement could be made to stagger the lunch breaks. Similarly, if there is a trolley service taking hot drinks or snacks to the workplaces, some staggering of the break periods is usually necessary.

Chapter 4

Refreshment facilities and mobile food vans

Mobile food vans

Mobile food vans are used to supply meals and snacks to workers on plantations, in large factories and mills and wherever the cafeteria is situated more than a reasonable walking distance from the workplace, for example, on large construction sites.

The design of the vans may vary. Some are simple flat-bed trucks carrying, for example, insulated food containers filled with hot beverages and hot and/or cold food. The degree of insulation required will depend on the climate - in some climates cold food may travel in uninsulated containers. More sophisticated vans will have self-contained refrigeration and heating facilities and may carry folding tables and chairs that can be set out under the shelter of one side of the van, which can be raised like an awning.

Whatever type of van is used, it should provide either a self-contained shelter or draw up alongside an existing shelter where the workers can sit and eat their food protected from the elements, whether in a cold or a hot climate. Since a van specially designed for this purpose is expensive, mobile food vans tend to be used mainly on very large sites and on rural sites where the workforce is isolated or dispersed. A contractor may run an enterprise which supplies meals to a number of isolated sites. The van will draw up at the site and someone on the ground will help to unload the food containers and then the van will drive on to the next site, coming back later in the day to collect empty containers, dishes and cups.

Food trolley trucks

Inside a factory or mill, special trolley trucks loaded with insulated containers of drinks and food are sometimes taken around behind a small tractor and left at a number of sites. Workers may serve themselves or a helper may be appointed to serve the food and beverages at each site. Later the tractor will pick up the trolley trucks and return them to the canteen.

Small food trolleys

Small trolleys fitted with hot water heaters and/or insulated containers for snacks and cups, etc., similar to those used in aircraft, can be pushed by hand around offices and workshops. Usually the cups will be washed after their return to the central canteen.

Where it is prohibited for food or drink to be consumed at the workplace, for example, where the work involves the use of toxic substances, the trolley can be parked outside the workshop or in a nearby passage. But it is important in such circumstances for the workers to be able to wash their hands before handling and eating food.

If the canteen is a long distance from the workshops or offices a tricycle or converted cycle rickshaw can be fitted to the trolley. Correct design of the trolley is important so that it is easy to push.

Trolleys of many different designs are available, but even in a small-to-medium-sized establishment the engineering staff should be able to design or build trolleys suitable for use in their own establishment.

Where a number of small establishments have a joint feeding service one or more trolleys will be essential for serving snacks. A decision as to who will service and repair the trolley should be made when it is first used.

Vending machines

Vending machines that supply hot or cold drinks, such as coffee, tea, soup or lemonade, and sometimes cold snacks and hot meals have become increasingly popular. They can replace the trolley which is wheeled round the workplace, and may remain in service when the mess room or canteen is not open.

A vending machine can be used in conjunction with a micro-wave oven to provide meals for night-shift workers. The correct amount of money is put in the slot and the frozen meal, already on a plate, is taken out. A token supplied on the plate operates the microwave oven alongside the vending machine. While this method of supplying meals has undoubted attractions in some situations, it has a number of disadvantages. In a country where labour for the canteen is available at reasonable wages, more traditional methods of cooking and serving are generally to be preferred.

The cost of leasing or buying a vending machine can be high. It may not be possible to have repairs done quickly if the terms of a lease specify that only the leasors' service staff may carry out repairs and maintenance. Routine cleaning, which is essential for health reasons, is a tedious task and workers are often reluctant to do it.

Summary

It should be emphasised that elaborate services to feed workers at the worksite are not necessarily the best service. The best is that which takes into account traditions and practice in the industry and the region, and provides food that is nutritious and tasty, acceptable to different groups, takes into account religious customs and is within the worker's budget.

Chapter 5

Local vendors

Facilities for vendors

In many countries hawkers, mobile food vendors and small corner cafes play an important role in feeding workers. If there is no canteen or mess room at their place of employment, workers who do not live with their families are virtually forced to buy food in the vicinity. Establishments may give a mobile food vendor permission to set up a van at fixed hours on the grounds of the enterprise to sell food direct to the workers, on condition that the prices are controlled by the establishment.

Some establishments provide facilities for one or more vendors to set up their stalls inside the enterprise grounds and will provide them with drinking water, free gas or electricity, sinks and hand-washing facilities. Two vendors are better than one as competition keeps up standards. Although vendors normally offer only a few dishes in which they have great expertise, they are usually skilled at adapting their recipes to suit the tastes of their customers. The prices the vendors charge should be controlled, and might be subsidised by the employer. The employer may place a few tables and chairs under a tree or other shelter near the stall, depending upon the climate - thus providing in fact a rudimentary mess room.

An employer or group of employers who provide a mess room may also allow food vendors or hawkers to sell food inside the factory gates at meal breaks. Where for security reasons hawkers and other food vendors are not allowed into the establishment they might be allowed to set up business for a short time, e.g. for two hours a day, outside the enterprise grounds. In some countries legislation lays down the distance the stall must be from an enterprise, school or hospital. If all the hawkers' customers are workers from the establishment, the management can justifiably check on the hygiene and prices. The employer can assist by ensuring the provision of clean water and providing garbage bins or sacks which can be stacked in the establishment's garbage enclosure.

Hygiene and health

Vendors do not always prepare food in hygienic conditions, and the employer would do well to look into this aspect before granting one permission to sell in the establishment. Vendors should be told that if they do not maintain such standards of hygiene they will lose the concession. The provision of clean running water will help the hawker to maintain such standards.

The factory doctor may be asked to give the vendor authorised to sell in the establishment a medical examination, at least annually. The enterprise's first-aid service should provide simple treatment for small cuts and burns, etc. If there is no health service covering the vendors (who are normally self-employed), the employer should ensure that the vendor is healthy in order to avoid diseases being transmitted to the workers and to ensure a reliable service. It is also advisable for senior members of staff and workers' representatives to make formal checks on the quality of the raw ingredients used by the hawkers and other food vendors.

An honest, clean and hygienic vendor who is provided with some facilities by the establishment, as described above, can provide a food service which will be very much appreciated by the workers and will help to maintain their fitness and health.

Some enterprises give their workers tickets or lunch vouchers which can be exchanged for a meal of a pre-determined value at a local restaurant or a vendor's stall. In some countries there is a central organisation which does all the book-keeping for such schemes.

Chapter 6

Canteens including inter-enterprise canteens

Introduction

Industrial canteens feeding large numbers of workers are operational throughout the world, and their number continues to grow. These canteens are set up to meet a vital need wherever men and women work, and in order to succeed they must be well managed.

A cafeteria that is almost empty does no one any good: the employer's money spent on premises, staff and fuel is largely wasted and the workers do not benefit from the nourishment provided nor from the psychological advantage of an enjoyable meal eaten in pleasant surroundings with friends and colleagues. A good cafeteria or industrial feeding service appreciated by workers is one of the intangible assets of a well-run establishment, be it a textile factory, foundry, department store, commercial firm, or whatever. Whether the canteen is small or is a large concern with well-trained, skilled staff under the supervision and leadership of a catering manager, the aim should always be to provide the best industrial feeding possible. Competent management is the key to the efficient running of an industrial feeding service.

Management responsibilities

The canteens of today fall into two groups: first, those set up to comply with the requirements of national laws and regulations; and secondly, those set up on a voluntary basis at the behest of management or the workers.

Where employers are obliged under national law to provide a mess room, canteen or other feeding service, they are ultimately responsible for the service, whether or not they assume direct management.

If the enterprise does assume direct and complete control, it may choose to run the canteen with the support of a strong and active canteen management committee comprising representatives of management and workers. Alternatively, the employer or a group of employers may appoint a contractor to run the service.

Since employers bear ultimate responsibility under national law, they can be expected to make periodic and sometimes unannounced visits to the kitchen, cafeteria and trolley service points during working hours, including night shifts. During these visits they will wish to satisfy themselves that good standards of safety and hygiene are being maintained, and that the food is eaten and enjoyed by the workers. They will also be interested in determining what proportion of the workers actually eat in the cafeteria or bring their own food to the mess room. If they observe that these numbers are low they will want to know the cause.

It may be that the time at which the cafeteria is open needs to be changed, or that the type of meals served may not be what the workers want. For example, if they have a long journey to work on foot or by bicycle they might prefer a breakfast before work rather than a midday or mid-shift meal. The meal break may not be long enough: a 20-minute break does not usually allow enough time for a worker to wash, walk to the cafeteria, queue up for a meal, eat it and then walk back to the workplace without an undue rush. A 30-45 minute break would be better. Such points as meal times, length of breaks and the type of meal served should first be discussed with the canteen management committee, and the workers' reactions should be sought through their representatives before any action is taken.

In cases where there is no legal obligation on the employer to provide an industrial feeding service, the service may be set up in one of several ways. The enterprise itself may run the canteen, or hand over the day-to-day running to a contractor, or use another method such as a co-operative of workers. A trade union may also contract to run the canteen either on behalf of the workers or on behalf of a joint employer/worker committee. Occasionally, a charitable or non-profit-making organisation may run a canteen or industrial feeding service (most likely in a depressed area) for one or more establishments. Whatever the arrangement, it should be remembered that the employer remains responsible under the law for at least the hygiene and safety of the premises. The wise and prudent employer will continue to maintain an interest in the feeding of workers and hence in the organisation, management and operation of the canteen services in the establishment. This latter point is the subject of extensive treatment in Part III.

Inter-enterprise canteens

Where several small firms have their premises located close to each other the idea of setting up an inter-enterprise canteen may be attractive from the economic point of view. It is an arrangement that can be very successful but also one that requires careful thought and planning before it is set up.

Apart from the economic aspects, practical considerations such as lack of land for individual canteens in crowded cities may oblige firms to consider setting up a canteen jointly. Industrial estates may be required by the terms of their charter to provide an industrial feeding service open to the workers in all the enterprises on the estate. Municipalities in industrial towns may run or subsidise cafeterias (sometimes called "Peoples' Restaurants") open to all workers and their families in the town.

Sometimes an employers' organisation may run an industrial feeding service for workers employed in member enterprises. This service may have a central kitchen from which cooked meals are distributed to individual enterprises who provide the cafeterias or dining halls. Workers' organisations may also run inter-enterprise feeding services, sometimes in co-operation with the employers.

The special requirements of inter-enterprise canteens are discussed in this chapter but the organisation and administration, methods of financing, physical facilities, equipment, maintenance and operation of a canteen whether single-enterprise or inter-enterprise are to be found in Part III, Chapters 8 - 12.

Canteen committees

The establishment of an inter-enterprise canteen requires due attention to detail if it is to succeed and run smoothly. As in the case of a canteen for a single enterprise, a planning committee should be created to set up the canteen and a management committee to run it. Managers of the various enterprises would sit together with workers' representatives on the planning committee, supported by professional staff (canteen manager, engineer, medical officer, accountant, etc.). It would, however, be necessary to determine how many enterprise owners, managers or their representatives would sit on the management committee, as well as the number of workers' representatives, technical and professional staff. Arrangements should also be made for substitute members.

Dining arrangements

While the inter-enterprise canteen will be similar in most respects to a canteen in a single enterprise with the equivalent number of workers, it may have to cater for persons with varying habits, customs and types of work. In its early stages the joint canteen may have to make modifications to its menus, hours of opening and type of service. The modifications required may be small but none the less important to workers, and consequently essential to the success of the canteen.

Central kitchen with peripheral cafeterias

Inter-enterprise canteens can also be based on a central kitchen where all food is cooked.

The cooked food is taken in insulated containers to cafeterias in different enterprises, where it is unloaded and served. However, some foods do not travel well and it may be necessary to have a "finishing kitchen" attached to each cafeteria in which certain foods can be prepared or finished.

This type of operation produces savings in skilled workforce needs and cooking fuel. It does however need very careful organisation and administration. Where there is a good supply of reasonably skilled workers there are fewer advantages in using a central kitchen. A variation on this arrangement is to set up a central food preparation organisation. With this arrangement vegetables and meat are prepared centrally, and prepared food is delivered each day (or several times a day) to the outlying canteens within the scheme.

A number of points need to be considered at the planning stage of such a scheme.

Menus

The type of menu and total number of meals served daily affect the size of the kitchen and the number of staff required. If the workers in the different establishments come from a number of ethnic or religious groups the menu will need to be varied to meet their needs.

The work carried out in different establishments may vary in its energy requirements, and this will affect menu planning. Workers on an open-air job, such as building a dam, heavy construction work or logging, will have main meal requirements quite different from those of a worker seated at a conveyor belt in a canning factory.

Distances

The distance between the central kitchen and the most distant cafeteria can affect the quality of the food. Many cooked foods deteriorate when transported unless special methods of handling and accurate temperature control are used. Airline catering provides an example of the routine use of very precise methods of control to minimise deterioration of food quality.

The need for such special handling and accurate temperature control has to be considered at the planning stage and may involve the installation of specialised equipment and additional staff training.

Transport

The suitability of the road between the central kitchen and each cafeteria should be considered. Certain foods will not withstand a journey on a winding or very uneven road, and the van will at least have to be so equipped as to prevent the food containers from being thrown from side to side.

Vehicles: fuel, maintenance, repair staff

The central kitchen should be allocated sufficient vehicles to carry the cooked food, as quickly as possible, to all the cafeterias. Possibly an outlying establishment will send a van to collect the food for its own cafeteria. Whatever system is used there should be adequate transport, including a back-up system of vehicles, to ensure that no workers go without a meal on account of a breakdown. Adequate supplies of fuel should always be available for the transport of meals.

Small frames on wheels, sometimes called "dollies", are useful to transport food containers to and from the van. Loading and unloading bays will be necessary.

Non-mechanised transport

Non-mechanised transport, such as cycle rickshaws and handcarts, may be more practical where distances between the central kitchen and cafeterias are short and the route lies along narrow alleys or lanes. Rickshaws come in various designs and one of the more suitable types for carrying food containers has the cycle mounted alongside the carriage. In areas where deep flooding occurs, the type with the cycle in front of a carriage mounted fairly high has certain advantages.

Handcarts also come in different designs. If suitable for carrying food containers the local design should be used. Otherwise a local cart-builder can be asked to make one to the desired pattern, and should be supplied with good working sketches.

There should be loading bays at the central kitchen and cafeterias that are at the appropriate height for any non-mechanised transport used.

Staffing and training

Skilled staff are even more important in an inter-enterprise than in a single enterprise canteen. Where skilled trained staff are not available locally, consideration should be given to arranging training before the canteen opens.

A trained or well-skilled supervisor engaged before the canteen is opened can be sent away for instruction as a trainer. Alternatively, a consultant with experience of training and opening new canteens can be employed on a short contract. Advisory services are run by hotel and catering management or home economics institutes in an growing number of countries and can be very helpful.

Work schedules

The work schedules of the enterprises served by the canteen will to a certain extent determine the times when the canteen is operative. Some may work a double shift and others may work round the clock. Where some or all of them work shifts, discussions should be held with the workers' representatives about the type of meals or snacks the workers prefer when working evening and night shifts. If they prefer to eat before or after the shift - at home with their family if possible - the canteen need only have a skeleton night staff to serve snacks or beverages.

When it is desired to serve an early morning snack to workers who live at a distance, the canteen will have to start work before the factories. Special transport facilities may be needed to bring canteen staff to work.

Chapter 7

Low-cost shops

Low-cost and fair-price shops

Low-cost shops (or stores) at which workers can buy food and sometimes clothing and household articles are sometimes set up in large enterprises. They may be called "fair-price shops" or "consumer societies". These shops are frequently run as co-operatives formed by members of a trade union, or by employees of an enterprise. Such shops are especially needed when new industrial units are set up in isolated locations.

Sometimes the employer will in the first instance establish and administer the shops later encouraging and helping the workers to set up a co-operative. The employer frequently provides premises at a very low rent or free of charge. In some regions of the world the major employers in certain industries traditionally provide such subsidised shopping for their employees and provide the administrative services or bulk purchasing, etc. In other regions unions tend to set up and run such shops.

In addition to premises and other facilities, the employer may be required to provide personal protective clothing. Such items may be sold to the worker at a reduced price or even supplied free. They include strengthened footwear, waterproof clothing for outdoor work, and trousers or slacks.

Certain matters should be considered and agreed upon at the setting-up stage. These include:

Choice of site

Convenience to customers - If the shop is to be located inside the factory it should be near the canteen/mess room or close to the main gate. If it is located outside the factory it should be near a bus stop or in the village where the majority of the workers live.

Security - If inside the enterprise grounds, the shop should be near the guardhouse or watchman's hut, or in a very well lit place and, if possible, linked to any intruder alarm system installed. If it is located outside the establishment, it should be in a well lit area, preferably a residential area.

Access for deliveries - The shop should be readily accessible to delivery vans, handcarts, etc., and have an unloading bay at a suitable height. For a small shop this could well be simply an old table with the legs raised or cut down to size. Where the shop building is not easily accessible, it may be possible to arrange for the establishment's own storage section to accept the delivery of goods.

Sales locations - good locations for the shop include:

- on the route to and from the workplaces,
- near bus stops used by the large numbers of workers,
- near to the customers' home.

A low-cost shop administered and subsidised by an establishment or a trade union will normally only be used by the employees or trade union members. A shop run by a consumer co-operative may be open to everyone, and not just its members. A co-operative may therefore prefer a shop site which, in addition to being near a bus stop and/or members' homes, is also close to a market, station or other busy place.

Type of building

Three types of building are commonly used:

- a kiosk - here customers are served outside through a window. The kiosk will be adjacent to a small store room.
- a shop where customers are served inside at a counter. The building should contain a small office and a store room. A veranda with some simple benches gives shelter and provides a meeting place for customers.
- a self-service shop where customers carry a basket round the shelves and pay a cashier as they leave.

Legal requirements

Local government planning permission may be needed before a building is erected or an old building converted into a shop.

Structural details

The building should be weatherproof.

It should have:

- Walls of concrete or fired brick. Sun-dried bricks or mud-and-wattle are not suitable. Corrugated iron should never be used for walls in a hot climate.
- A roof of a suitable material such as corrugated iron, slates or concrete. Flattened kerosene tins are initially cheap but have to be renewed frequently. Thatch is not suitable because of the fire risk.
- Small ventilators, covered with fly-screening, high up on two oppposite walls (for preference facing north and south in hot climates).
- Gutters - in wet climates there should be gutters to lead rain water from the roof into a drum or barrel.
- Drains - in wet climates there should be drains or ditches lined with stones round the entrances to the shop to prevent water collecting in pools.
- Doors - two only for a small to medium sized shop - one for customers and one for staff and goods delivery.

  Both doors should be able to be bolted from inside and should have strong locks. At night the customers' entrance may be bolted from inside as well as locked. By day the goods entrance may be bolted from the inside but not locked (because of the need of escape in case of fire) when the customers' entrance is open. There should be strict regulations about who is permitted to handle keys and where the spare set of keys is kept for emergency use.

- Windows - the guide-lines in the chapter on canteens for the siting and screening of windows in cafeterias can be followed. There should also be bars and strong shutters for when the shop is closed.

- Lighting - if artifical light is needed in a shop located in a village it may be necessary to use kerosene (paraffin) lamps. There should be simple but strict instructions as to who may handle the lamps and where they may be hung. A fire blanket and bucket of clean sand are minimum fire-fighting requirements when such lamps are used.

- Heating - whenever possible the use of artificial heating should be avoided. Whatever form of heating is used must be clean and as safe as possible. Kerosene, wood, coal or charcoal heaters present a fire risk unless they conform to recognised safety standards and are well maintained. If any form of heating is used, appropriate fire extinguishers, fire blankets and sand should also be provided.

- Sanitary accomodation - toilets (WCs) and washing facilities. If the shop is owned, administered or subsidised by the enterprise it may be possible to use the enterprise's facilities. A co-operative outside the enterprise should follow local government regulations. Hygiene in the shop is improved if there are adequate toilet and washing facilities. The very minimum provision would be a wash stand or basin at the back or outside the staff entrance with a drum or barrel of water and a dipper.

- Goods storage - the guide-lines in the chapter on canteens should be followed as far as possible when planning the storeroom of the shop. The establishment's own store-keeper could be asked for advice and help and invited to join to the shop's management committee.

Training in the establishment and the administration of a low-cost shop

In some countries, government and semi-governmental organisations give advice on setting up small businesses, and they should be asked for advice at the planning stages of establishing a low-cost or fair-price shop. Where government regulations cover the setting up of such shops, these must be followed. Sometimes government aid is available to industries in certain categories for the establishment of such shops.

Trade unions may have advisers able to train the staff of a low-cost or fair-price shop. Co-operative movements, where they exist, are usually able to advise on most aspects of the creation of a consumer co-operative, whether it is in an industrial establishment or outside. For further information, contact the national co-operative society.

A series of illustrated booklets produced by MATCOM (a joint project of the International Labour Office and the Swedish International Development Authority) for training the staff of small consumer co-operatives are useful to any organisation, whether private, public or co-operative, about to set up a low-cost or fair-price shop.

The MATCOM booklets are listed in the bibliographical section.

Merchandise

The type of goods a low-cost or fair-price shop should sell will depend on the needs of the workers and on local conditions, such as existing markets and shops. The location of the establishment or group of establishments will also have an influence.

In some remote areas it has been found that mail-order shopping (with the shop acting as an agent of the mail-order company) for clothes and household items is appreciated. In other areas, where basic foodstuffs are readily available, the workers may appreciate the chance to buy special foods, e.g. tinned or packaged foods for festive occasions and other foods which are unobtainable or expensive to buy locally.

Shiftworkers

The opening hours of the low-cost shop should be arranged to suit the working day in the establishment or group of establishments. Where shifts are worked, this may mean a late-night or early-morning opening once a week.

Home-grown vegetables and other produce

It is common in some countries to sell home-grown produce, eggs and home-baked biscuits, etc., in co-operative stores or from a temporary stall on boxes laid out by the factory gates or just outside a mess room.

This has the double advantage of increasing the family income of the producer as well as supplying fresh, good quality food to other workers at a price they can afford.

PART III: ORGANISATION, MANAGEMENT AND FINANCING OF CANTEENS

Chapter 8

Organisation and administration

Introduction

Chapter 6 discussed the need to determine what type of canteen would best suit an enterprise or group of enterprises and in this section of the manual, attention will be drawn to:

- the desirability of involving all levels of the enterprise workforce, management and workers in the planning and running of the canteen;
- the need for careful choice of a site for the canteen;
- the choice of equipment for the canteen;
- budgeting for servicing and maintenance of equipment and buildings;
- the importance of medical surveillance of the canteen staff;
- the importance of housekeeping and proper safety precautions to ensure a safe place of work;
- the desirability of selecting skilled catering staff; and
- training and retraining of canteen staff.

Workers' participation in running a canteen

The establishment of a canteen planning and management committee is always desirable, whether or not one is legally required. Joint canteen planning and management committees where workers are also represented can facilitate the accurate reflection of workers' dietary preferences and the planning and efficient operation of feeding services. If a trade union or workers' cooperative is running the canteen, they could also have a committee on which the users would have their representatives in order to ensure that customers are served food which tastes good, conforms to religious dietary practices and is cooked and served hygienically.

Generally, there will be a need for a short training course, run by a trade union, a co-operative training centre or a technical institute, on the art of active constructive participation in the work of a committee. Evening courses of this nature may be provided in restaurant or hotel management institutes or colleges.

A canteen planning and management committee should involve the medical officer, the canteen manager and the engineer. The role of the committee secretary is important, and the person elected to carry out this task should be given adequate assistance for typing, photocopying or duplicating the minutes, agendas, etc. and also be allowed to use the establishment's central service for their distribution.

While committee meetings must be held regularly, complaints about food or service are best dealt with as soon as they are made. A report on the complaint, its examination and the proposed solution should then be submitted to the committee at its next meeting.

Staff

The most valuable asset of any industrial feeding services is its staff, and its success depends largely on them. The staff should be carefully selected, be well trained and be given good working conditions and wages. It is important that the staff feel that they are part of the whole workforce in an enterprise. Therefore, ways and means of involving them in decision making and in social activities should be designed.

Staff titles and duties

The head chef of a large establishment has mainly administrative duties - he or she organises the work, compiles menus, orders food, oversees the budget, engages staff, supervises the work of the kitchen (especially at meal times), advises on the purchase of equipment, and may be responsible for the stores and the care of dishes, cutlery, etc. In a small establishment this role is filled by a cook/supervisor.

The sous-chef or second chef of a large establishment is the head chef's deputy, and acts for the head chef when he or she is absent, assists in supervising staff, and may have responsibility for a section of the kitchen. This role in a small establishment is filled by a cook.

The commis chefs in large establishments are assistant cooks with a considerable degree of responsibility. In a small establishment the assistant cook understudies the cook and is responsible for part of the work, such as the cooking of certain types of food.

The general assistant in a small kitchen helps the cook and assistant cook with various parts of their work, but will normally have very limited responsibility. However, in some kitchens each general assistant is given responsibility for a section of the work - one being responsible, say, for the preparation of vegetables and for the tea trolley at coffee/tea breaks, and another for the operation of the dish-washing machine and the cleanliness of the hot-plate and serving counter in the cafeteria, while the two may share responsibility for cleaning the kitchen floor on alternate days.

Catering or canteen manager

In large organisations where the size of the operation warrants it, a catering manager will be appointed. He or she will have greater responsibility for financial matters than a cook/supervisor and should have received training and some experience in supervision and management. The manager's rank in the establishment's hierarchy will be the same as that of the managers of equivalent-sized departments, whereas the cook/supervisor's rank will normally be more equivalent to that of a foreman or department supervisor. The catering or canteen manager may have an office in the administrative block near the other managers, while the cook/supervisor who deals with the day-to-day administration of the canteen will have an office in the canteen itself.

Nutritionists and dietitians

The training of nutritionists and dietitians varies considerably from country to country, both in the subjects studied and the length of training. In consequence the use of these titles for persons with different skills and levels of knowledge has led to some confusion. In particular, the training may or may not cover aspects of industrial feeding or the administration and management of large-scale catering operations.

Government regulations sometimes prescribe that a dietitian or nutritionist must be appointed where meals beyond a certain number are served daily in an industrial canteen. Such regulations are usually found in countries where dietitians/nutritionists receive special training in the administration of large-scale catering.

In some countries, dietitians and nutritionists may specialise in clinical situations. Such specialists are not necessarily the most suitable persons to employ in an industrial feeding service, although they can advise on the nutritional content of the menus and help to plan modifications for workers with diabetes, coeliac disease, etc.

Size of staff

The number of staff and the work they carry out in an industrial feeding service will vary according to the size of establishment. It will also vary according to the type of meals or snacks served, on whether labour-saving equipment is used in the kitchen, on geographical location and climate, on the staff's skills, on shift arrangements and work schedule, and on staff composition.

Consequently the suggestions given as to staffing levels should be regarded as approximations only, since each canteen has individual requirements. When an industrial feeding service is initially set up it is useful to provide in the budget for possible increases in the canteen staff after the first year of operation so as to allow for a growth in its use by the workers.

It is assumed that the daily cleaning of the cafeteria and adjacent premises will be carried out by the general cleaning services of the establishment. Kitchen cleaning will normally be done by the kitchen staff, although which members of the staff do the cleaning varies according to local custom.

The table below gives approximate staff requirements for a kitchen providing one main meal daily and two coffee/tea breaks per day with a trolley service on a five-day or five-and-a-half-day week.

Number of meals prepared	Cook/ supervisor	Cook	Assistant cook	General assistants
25 meals & 75 coffee/ tea breaks	1	-	1	1 or 2 part-time
50 meals & 150 coffee/ tea breaks	1	-	1	2 full-time & 2 part-time
100 meals & 300 coffee/ tea breaks	1	1	1	3 full-time & 2 part-time

In these arrangements the cook/supervisor would hold overall responsibility for running the canteen, would have some clerical assistance for store-keeping, ordering and record-keeping and would also carry out some of the cooking and be in charge of the meal and trolley service.

The cook would be responsible for the greater part of the cooking and would supervise the work of the assistant cook and general assistant, and would deputise for the cook/supervisor during short absences.

The assistant cook would prepare certain dishes as instructed by the cook/supervisor or cook, and would directly supervise the preparation of vegetables, rice, noodles, pasta, etc. by a general assistant.

The general assistant(s) would be responsible for cleaning vegetables, operating various food processing devices, such as slicers and potato peelers, operating the dishwashing machine, cleaning cafeteria tables, taking trolleys to the workplaces and general cleaning of the kitchen.

Engaging new workers

Medical examinations. The selection of canteen staff is important. One aspect of the selection procedure should include a medical examination, given before an applicant is offered a job in a canteen and at regular intervals thereafter. Eyesight should be tested, and glasses worn at work if considered necessary. The International Labour Organisation's Occupational Health Services Convention, 1985 (No. 161) provides a useful guide to desirable standards for pre-assignment medical examinations, health assessment and surveillance of food handlers and is of interest to industrial medical officers, nurses and personnel officers.

Induction of new canteen workers. On-the-job training of canteen staff must start from the day they first report for duty. The first few days at work should be used to introduce new workers to their colleagues in the kitchen and cafeteria, and to the administration of the enterprise for completion of the necessary formalities. The induction process should cover explanations of working arrangements, the staff hierarchy, hours of work and holiday arrangements; the issue of protective working clothes and allocation of a locker in the changing room. Very important in these first two days is instruction by the supervisor or senior cook in the standards of hygiene and cleanliness required of all staff.

The supervisor or a person delegated by them should inform the person in charge of first-aid/fire-fighting that a new worker has been engaged and arrange for them to be included in the next safety class or demonstration. The new worker should be shown the location of fire extinguishers and blankets for the canteen and informed of the action to be taken in case of fire.

Part-time staff. Part-time staff can be of value in a canteen service. Local staff may be willing to work every day for a short period, even if only for two hours during the busy period when the main meal is served (provided that such an arrangement is permitted under the Labour Regulations). The supervisor should keep a list of part-time workers who could be engaged to cover staff shortages.

Shift work and night duty. Where the canteen provides meals for workers on shifts it will be necessary for several of the staff to work rotating shifts, or for separate teams to work evening and night shifts. The system chosen will depend upon the locality and local custom. It may be possible in a two-shift system to employ a team of part-time workers for the evening shift. Night shift in a canteen can be rather lonely and, if possible, there should always be two people on duty.

Where a rotating shift system is operated in an establishment, it is a good idea for the canteen staff that also work shifts to follow the same system, i.e. cook A and cafeteria assistant B will always be on duty with, for example, No. 1 shift, while cook C and cafeteria assistant D will always be on duty with No. 2 shift, and so on. In this way the cook and cafeteria assistant come to know the co-workers on their own shift.

Contractor's staff

Where a contractor is employed, due consideration should be given before the contract is signed to the type of staff to employ, the standard of protective clothing to be provided and the conditions of work and wages.

Good contractors have high standards in these matters, but it is always wise to check. The contractor is engaged primarily to feed the establishment's workers in a safe and hygienic manner, and the quality of service provided should be monitored regularly.

Industrial canteens in a developing country will be competing for the available skilled staff with large hotels that can pay higher wages. The canteen may then have to accept what might be termed "second-best" staff. However, given good supervision and management, and encouragement to attend training courses not just in cooking skills but in accounting, stock-keeping, and management development, these workers may become very competent.

Training

In catering it is common to combine three types of training, (a) theoretical and practical training at a school, college or institute, (b) practical in-service training, either in the establishment where the trainee is employed or apprenticed, or in another establishment approved by the training institute for such practical training and (c) training visits to establishments doing similar work. When the place of employment is far from a training institute, the theoretical side of the training may be obtained through a correspondence course and the student will take externally monitored practical examinations and theoretical examinations.

Catering is still a profession in which it is possible for an ambitious person to start at or near the bottom of the profession and climb to the top. It is also a profession in which women hold managerial posts, especially in industrial catering. A person with such ambitions will choose to work in establishments with a good reputation for in-service training and will move from establishment to establishment in order to gain varied experience.

It is unlikely that a small to medium-sized industrial catering service in a developing country will be able to attract and recruit such skilled staff. More probably, they will have to recruit unskilled or semi-skilled staff from local villages and towns with little or no formal training. Much can be done, even in small canteens, to train and upgrade staff and by so doing to improve the quality of the food and service. A supervisor or manager can be sent to a supervisory development course, a training-within-industry or a trainer-skills course. Cooks who obviously have the ability to communicate and teach can be sent on a course to upgrade their technical skills, and on their return they can teach others what they have learnt. Courses in management development are just as important for catering staff of equivalent rank as for their colleagues on the shop floor. Similarly, courses in basic book-keeping and the use of calculating machines and computers can all help to improve the efficiency of the canteen. Catering staff should attend any first aid, fire fighting and civil defence classes and exercises organised by the establishment. A qualification in first aid is a professional asset for canteen supervisors. Adapting the type of training to the worker is worthwhile and even if some of the staff are attracted away to another establishment by higher wages, it will be found that recruitment is easier once an establishment has a good reputation for training.

Even if an establishment is too small to have a training section, there may well be a training adviser attached to the head office of an employers' group for the industry, who can advise on general training schemes for the canteen staff. In some countries, colleges or institutes teaching catering subjects are scattered. It may be difficult to attend training courses which are at a distance from the workplace, and staff may benefit more from attending a short intensive course lasting one or two weeks on a residential basis rather than undertaking a long and tiring journey to attend a weekly evening course over a whole year.

Countries developing a tourist industry often have excellent hotel, restaurant and catering institutes or colleges that train potential hotel and tourism staff. While the majority of the students will have little or no interest in industrial catering, the staff of these colleges will often be willing to give advice to industrial caterers on technical matters. It might even be feasible on the request of a group of employers or a trade union for special short courses to be arranged for senior canteen staff.

Small or medium-sized establishments may find an informal source of advice on training and technical matters in a large firm for whom they do contract work. This larger firm may well have a catering manager who would be happy to put his or her professional knowledge at the service of the smaller firm.

Modular System of Training

The ILO* has prepared a system of modular training for the use of those engaged in training hotel and restaurant staff. Although primarily intended for use by instructors in developing countries, it is also used by instructors in the industrialised countries.

The introduction to the handbook: Tasks to Jobs - Developing a Modular System of Training for Hotel Occupations states:

"...The idea in its simplest form is based on the identification of tasks in a given sphere of operations and the selection and clustering of numbers of tasks to form jobs. However, for training purposes, the gap between the identified tasks and the job performance is wide and the needs of trainers and trainees require supplementary information...

...a system which, without regard for consideration of time would lead to the acquisition of employable skills."

---

* Hotel and Tourism Branch.

Examples from Operating self-service counters, Operating dish-washing machines and Supervising are given below.

K/04 Operating dishwashing machine

Task elements:

K/04.01	Checking that the machine is ready to operate
K/04.02	Removing food particles from the items to be washed
K/04.03	Soaking
K/04.04	Loading the machine
K/04.05	Stacking and storing glassware, crockery and cutlery
K/04.06	Emptying and cleaning the machine

Skills, knowledge and attitudes:

Ability to:
- comprehend, read, write, calculate and communicate,
- adjust dishwashing machine for operation,
- operate various kinds of dishwashing machines,
- dismantle and clean dishwashing machines,
- recognise and report malfunctions
- keep working areas safe, clean and tidy,
- maintain tools and equipment in good working order.

Knowledge of:
- techniques for washing glassware, crockery and cutlery,
- structure and components of various dishwashing machines,
- soaking and dishwashing agents,
- damage prevention,
- stacking and storing of glassware, crockery and cutlery,
- elementary maintenance of dishwashing machines,
- safety and hygiene standards.

Attitudes:
- care,
- tidiness,
- sense of responsibility.

R/13 Operating self-service counter

Task elements:

R/13.01	Taking stock
R/13.02	Obtaining supplies
R/13.03	Arranging commodities
R/13.04	Replacing commodities as required
R/13.05	Arranging self-service equipment
R/13.06	Replacing as required
R/13.07	Keeping the counter and its surroundings clean (See also task C/01)
R/13.08	Maintaining a smooth work-flow

Skills, knowledge and attitudes:

Ability to:
- identify goods,
- obtain supplies,
- handle goods correctly,

- use cleaning utensils and agents (see also task C/01),
- operate machinery,
- identify and report malfunctions.

Knowledge of:
- counter plan,
- storage of goods and equipment,
- hygiene and safety,
- relevant legislation,
- psychology of the customer,
- merchandising and display,
- work-flow systems,
- departments of the establishment.

Attitudes :
- care,
- neatness,
- integrity.

C/07 Supervising (see also tasks F/36, R/42, B/19, K/43, and H/33)

Task elements:

C/07.01	Drawing up job specifications
C/07.02	Assessing qualities and qualifications of team
C/07.03	Apportioning tasks
C/07.04	Delegating responsibility
C/07.05	Advising and helping when necessary
C/07.06	Reprimanding when necessary
C/07.07	Inducting new members of the staff
C/07.08	Ensuring adherence to work norms
C/07.09	Controlling behaviour/appearance/dress
C/07.10	Encouraging
C/07.11	Motivating
C/07.12	Recommending for promotion
C/07.13	Recommending for dismissal if necessary
C/07.14	Giving personal example
C/07.15	Taking personal interest
C/07.16	Interviewing
C/07.17	Hiring
C/07.18	Co-operating with other departments

Skills, knowledge and attitudes:

Ability to:
- lead,
- observe, assess and quantify qualities and qualifications of team members, individually and collectively,
- add individuals into effective team force,
- designate and allocate tasks fairly,
- impose discipline with justice and good judgement,
- display personal capability,
- recognise human weakness and encourage honest effort.

Knowledge of:
- human relations,
- interviewing techniques,
- labour legislation,
- establishment policy,

- principles of supervision,
- principles of leadership,
- career opportunities,
- conditions of employment,
- dress and behaviour patterns,
- job specifications.

Attitudes:
- dignity,
- justice,
- interest.

Training films and posters

Films are a valuable means of teaching, particularly in the fields of safety, food handling and hygiene.

Suitably chosen posters, changed regularly, can improve standards of safety and hygiene. However, posters designed for one country are not always successful in another, and a local artist, or someone on the staff of the establishment, can often design suitable posters once their purpose is made clear. Posters lose their impact if left up too long. They should be changed frequently, for example on the first day of each month. The same poster can be shown again after a break.

Financial management

The importance of financial management, including accurate stock control, elimination of waste and accurate budgeting has already been stressed in earlier chapters. In a small canteen the staff will generally have no experience of record-keeping, and some clerical help may be necessary. Regular help by the office staff may be more satisfactory than trying to find a cook or supervisor with an aptitude for clerical work.

Computers

Calculating machines and microcomputers are increasingly used in catering. They are of value in stock control, costing of recipes and ordering foods.

A medium-sized establishment may already have a microcomputer in use in the accounts department which the canteen manager or the store-keeper may be able to use to simplify their work. Properly used, it can bring considerable savings by pinpointing areas of waste, simplifying the calculation of meal prices when the prices of raw ingredients vary, speeding up stock-taking, and making advance ordering possible without much risk of overstocking.

Once a basic recipe has been entered into a computer it is a simple matter to add the number of portions to be prepared and to have the computer print out all the ingredients, quantities, oven or steamer space required, and the cost per portion.

Senior canteen staff in large industrial enterprises should be encouraged to attend any introductory courses on computers available in the training department or at a nearby college, for example.

Pricing and budgeting

A newly graduated catering manager or supervisor will be trained to maintain a close control over costs, but in a small industrial canteen it may not be possible to obtain the services of such qualified staff. However, close control of costs, elimination of waste, prevention and early detection of petty pilfering, and wise purchasing to obtain the best value for money are all an integral part of the successful operation of any department of an industrial establishment, be it the tool-room, foundry, assembly line or canteen.

Recipe cards

Recipe cards should be prepared for each dish on the canteen menu for the use not only of the cook but also of the person who orders the raw ingredients. These cards, which should be updated when necessary, will indicate the quantity of each ingredient required for a given number of portions, the various steps for preparation and cooking, and the times required. In canteens catering for different ethnic groups the card should indicate the groups to which the dish appeals most, and whether it is forbidden by any particular religious or tribal group. If desirable, pictorial recipe cards can be made with drawings showing how many tins, cups, spoonfuls, etc. are required, the type of heat being illustrated by stylised flames or a picture of an oven dial.

When a canteen is upgrading the quality of its food it will be useful to note down the recipes that have been in use over a period of time. The following example of a card used for stock recipes can be modified to suit the individual canteen's requirements.

Name of Canteen......................Recipe for......................

Qty. Ingredients.

Type of dish....................

Number of portions.............

Time for preparation...........

Time for cooking...............

Cook in steamer ...
oven ...
tilting fryer ...
grill ...
fryer ...
hob ...
tilting kettle ...
microwave ...

Serve in.......................

Remarks........................
....

Method 1.

2.

3.

4.

5.

Quantities can be shown in pictures of standard cups and spoons.

Stock keeping

The person responsible for ordering, receiving and issuing stores is a key person in the control of costs in the canteen. In a small canteen the supervisor may also control the stores and someone from the accounts department might help with record-keeping. Examples of some typical forms used are given elsewhere.*

In any case, each time goods are received or issued the appropriate entries should be made on the necessary stores ledger sheets and bin cards. In this way the balance on the bin card should always be the same as the balance shown on the stores ledger sheet.

All requisitions must be handed to the storekeeper in time to allow the ordering and delivery of the goods on the appropriate day. Different coloured requisitions may be used for the various departments if desired.

(a) Departmental requisition book: One of these books should be issued to each department in the catering establishment which finds it necessary to draw goods from the store. These books can either be of different colours or have serial numbers denoting the department to which they belong. Every time goods have to be drawn from the store a requisition must be filled out and signed by the necessary head of department. This applies whether one item or twenty items are needed from the store. When issuing the goods, the storekeeper will check them against the requisition and tick them off, and at the same time fill in the cost of each item. In this way the total expenditure over a given period for a certain department can be quickly found. The following details are found on the requisition sheet:

1. Serial number
2. Name of department
3. Date
4. Description of goods required
5. Quantity of goods required
6. Units used
7. Price per unit
8. Issue, if different
9. Quantity of goods issued
10. Unit
11. Price per unit
12. Cash column
13. Signature

(b) Order book: This is in duplicate and has to be completed by the storekeeper each time he or she wishes to have goods delivered. Whenever goods are ordered, an order sheet must be filled in and sent to the supplier. On receipt of the goods they should be checked against both the delivery note and the duplicate order sheet. All order sheets must be signed by the storekeeper. Details found on an order sheet are as follows:

1. Name and address of catering establishment
2. Name and address of supplier
3. Serial number of order sheet
4. Quantity of goods
5. Description of goods to be ordered
6. Date
7. Signature
8. Date of delivery, if specific day required

(c) It is also useful to check the stock periodically, for example, once every three months. Stock sheets used for this purpose show the following details:

---

* See, for example, R. Kinton and V. Ceserani: The Theory of Catering, fifth edition, London, Edward Arnold, 1985.

Daily Stores Issues Sheet

DAILY STORES ISSUES SHEET

Commodity	Unit	Stock in hand	Monday In	Monday Out	Tuesday In	Tuesday Out	Wednesday In	Wednesday Out	Thursday In	Thursday Out	Friday In	Friday Out	Total purchases	Total issues	Total stock
Butter	kg	27		2						3				5	22
Flour	sacks	2		1			1						1	1	2
Oil	litres	8		1						1/2				1 1/2	6 1/2
Spices	30g packets	8		4			8						8	4	12
Lentils	kg	30		6						3				9	21

CANTEEN .... Week ending .... No. meals served .... Cost per meal ....

Commodity	In hand B/F	Stock received during week						Stock used during week									In hand C/F
		M	T	W	Th	F	Total	M	T	W	Th	F	S	Total	At*	Cost*	
Salt																	
Apples, dried apricots, etc.																	
Baking powder																	
Chickpeas																	

* The cost of stock used can also be checked by using two extra columns.

Canteens and food services in industry:
A manual

Commercial documents

Essential parts of the control system of any catering establishment are delivery notes, invoices, credit notes and statements.

Delivery notes are sent with goods supplied as a means of checking that everything ordered has been delivered. The delivery note should also be checked against the duplicate order sheet.

Invoices are bills sent to clients, setting out the cost of goods supplied or services rendered. An invoice should be sent on the day the goods are despatched or the services rendered, or as soon as possible afterwards. At least one copy of each invoice is made, and used for posting up the books of account, stock records and so on.

Invoices contain the following information:

(a) The name, address, telephone numbers, etc. (as a printed heading) of the firm supplying the goods or services.

(b) The name and address of the firm to which the goods or services have been supplied.

(c) The title INVOICE.

(d) The date on which the goods or services were supplied.

(e) Particulars of the goods or services supplied together with the prices.

(f) A note concerning the terms of settlement.

Credit notes are advices to clients, setting out allowances made for goods returned or adjustments made through errors of overcharging on invoices. They should also be issued when chargeable containers such as crates, boxes or sacks are returned. Credit notes are exactly the same in form as invoices except that the word CREDIT NOTE appears in place of the word INVOICE. The following terms are important and may need clarification:

Statements are summaries of all invoices and credit notes sent to clients during the previous accounting period, usually one month.

Cash discount is a discount allowed in consideration of prompt payment.

Trade discount is discount allowed by one trader to another.

Gross price is the price of an article before discount has been deducted.

Net price is the price after discount has been deducted; or in some cases a price on which no discount will be allowed.

Budgeting

In addition to the cost of food, labour, fuel, depreciation of capital equipment, local taxes, and social security payments, the industrial feeding service budget must also make provision for:

Cleaning supplies and the replacement of cleaning equipment, such as brushes and mops, garbage sacks, hoses and buckets. Provision and replacement of protective and hygienic clothing for workers;

Laundering of protective clothing, whether by an outside laundry or in the form of a laundry allowance to staff;

Laundering of kitchen cloths, etc. (some canteens prefer to employ someone to do this on their premises and provide detergent and other supplies, an iron and possibly a washing machine);

Servicing and maintenance of kitchen equipment and the regular cleaning of walls and ceilings (this may go on to the engineer's budget);

Any overtime worked by kitchen and cafeteria staff; possibly also transport home for canteen workers who work overtime;

Urgent repairs: although regular servicing and maintenance of equipment minimises the risk of breakdown, there should be some provision in the budget for urgent repairs or for paying rental for a replacement machine if this is necessary.

Maintenance of vending machines

A reluctance on the part of staff to clean vending machines has been noted by manufacturers. Newer models can be flushed automatically twice daily with hot water, thus reducing the chore of cleaning to an acceptable minimum. Models supplying chilled water now have their own integral water purifying filters.

A vending machine out of service for several days, because no service staff are available in the vicinity, is unsatisfactory. This question should therefore be considered before vending machines are installed. If there is no agent sufficiently near, then the establishment's own engineers or maintenance staff should be given instruction on maintenance by the supplier.

One of the factors in gaining acceptance of a vending machine in an establishment is the quality of the food or drink that it serves. The basic quality of the food used varies from supplier to supplier and, before installation is considered, an evaluation should be made by members of the canteen management committee and other representatives of workers. In particular the quality of the tea or coffee used is critical, and it must suit the tastes of those who have to drink it.

Whether a vending machine is appreciated by the night shift or not also depends on the atmosphere of the canteen or mess room where it is installed and the presence of a pleasant steward or attendant.

"Cashless cards"

A recent development in automatic vending is the result of the invention of plastic cards which can hold a quantity of information in a very small space.

Among the many uses of these cards in industry is the possibility of allowing passage into a cafeteria providing meals free or at a fixed rate. It can also be used in vending machines to obtain meals or drinks.

Such cards may also have a cash value, which can be repeatedly "re-charged" or "topped up" by putting money into a machine.

It may also be possible to re-charge or top up the card by direct debit from a salary account.

Inside the cafeteria the card can be used to pay for a meal by inserting it into a slot at the side of the cash register. A display will then show the cost of the meal, the present value of the card and, after the transaction, the new value of the card.

An advantage of such a "cashless card" system is that it can be so programmed that different grades of staff pay different amounts for the same food, i.e. a labourer will only pay a token cost for a meal while the executive will pay the full economic cost. It is also possible to programme the system so that a card holder can obtain two or more drinks free of charge per day.

Where a small or medium-sized undertaking is considering introducing a card system to control access to high security areas of the establishment, this extended use of "cashless cards" for catering offers an added advantage.

The card system can also be used with small portable units that accept the cards to provide a hot or cold beverage. Such a unit may be carried on the traditional trolley taken around workplaces during working hours.

Chapter 9

Methods of financing

Food costs: The elements and implications

Firms and establishments differ in size, location and activity. Consequently the methods of organising and financing food services also vary among industries and enterprises.

As pointed out earlier, a feeding service may be a breakfast given to workers who have long journeys before they start work, a snack and beverage in the middle of a shift, a messroom where workers can eat their own food and make hot drinks away from the workplace, a canteen with cafeteria or dining hall serving full meals, and so on. Each of these involves a cost. Each also involves different options not only as regards forms of providing access to food but also as regards the financial burden of doing so.

The cost of providing food for workers consists of expenditure associated with the provision of (a) land, (b) buildings, (c) equipment, (d) services, e.g. electricity, water, etc., (e) maintenance, (f) foodstuffs and (g) canteen staff. Installation costs vary from one canteen to another depending in part on the size, the type of service, the quality of equipment or the type of interior decoration. The cost of raw materials will depend on availability of food supplies and the quality and quantity needed. Staffing costs vary according to local wage rates, number of workers employed and national laws and practices concerning social security benefits. All of these are important in determining the cost of the meal to the worker which can be critical in the utilisation of a feeding service.*

The cost of providing feeding services through formal structures, such as canteens, can be high and burdensome for employers. Some form of cost sharing between employers and workers is therefore often necessary and desirable. The International Labour Organisation's Welfare Facilities Recommendation, 1956 (No. 102), provides some general guide-lines on this. For example, the Recommendation suggests financing by the employer of expenditure for constructing, renting or providing premises and equipment and for maintenance and overheads on the one hand, and payment for meals and other food supplies by the workers using the facilities on the other. The provision of such facilities as canteens and the financing by employers even of installation costs can nevertheless be difficult for many. They may constitute a heavy financial charge,

---

* Many firms have found that if the midday meal is free or is costed low compared to say the lowest paid worker's daily wage, the majority of workers will eat it. A sliding scale of charges for meals with lower paid workers paying less than the higher paid workers for the same meal is sometimes advocated. But this can be a complicated operation and can only be achieved where there are either (i) separate dining rooms for different grades of staff, or (ii) a system whereby meals are charged on a credit basis and deducted from wages or salary (which is not always permitted under national regulations) or (iii) where "cashless cards" are used for vending machines and at cash tills. Sometimes the enterprise and the workers through their representatives reach an agreement on the type and price of meals or snacks to be served. In some countries, collective agreements link the cost of meals sold in a canteen to the cost of the raw foodstuffs plus a fixed percentage, say 10 per cent (to allow for the cost of transport, handling, etc.) of services.

FINANCING OF FEEDING AND RECREATION FACILITIES

25. While the financing of the feeding and recreation facilities provided may be exercised in different ways in accordance with the customs of the country or locality concerned or with arrangements under which special bodies are entrusted with overall responsibility for welfare facilities, the following are some of the forms of financing that competent authorities, employers and workers should take into account:

(a) in respect of feeding facilities -

(i) financing by the employer of expenditure for constructing, renting or otherwise providing the premises for feeding facilities together with the necessary equipment and furnishings and for continuing overheads and maintenance, including heating, lighting and cleaning, rates and taxes, insurance and upkeep of premises, equipment and furnishings;

(ii) payment for meals and other food supplied by the workers using the facilities;

(iii) financing of expenditure for wages and insurance of food service personnel, either by the employer or by the workers through payment for meals and other food supplied;

...

26. In the economically underdeveloped countries, in the absence of other legal obligations concerning welfare facilities, such facilities may be financed through welfare funds maintained by contributions fixed by the competent authorities and administered by committees with equal representation of employers and workers.

27. (1) Where meals and other food supplies are made available to the workers directly by the employer, their prices should be reasonable and they should be provided without profit to the employer; any possible financial surplus resulting from the sale should be paid into a fund or special account and used, according to circumstances, either to offset losses or to improve the facilities made available to the workers.

(2) Where meals and other food supplies are made available to the workers by a caterer or contractor, their prices should be reasonable and they should be provided without profit to the employer.

(3) Where the facilities in question are provided by virtue of collective agreements or by special agreements within undertakings, the fund provided for in subparagraph (1) should be administered either by a joint body or by the workers.

28. (1) In no case should a worker be required to contribute towards the cost of welfare facilities that he does not wish to use personally.

(2) In cases where workers have to pay for welfare facilities, payment by instalment or delay in payment should not be permitted.

Welfare Facilities Recommendation, 1956 (No. 102)

---

particularly for small and medium-sized enterprises whose financial means are limited but which in many countries provide employment for a significant proportion of the labour force. Hence, different forms of organising and financing feeding services are and may be adopted in different countries at different levels of development.

Methods of organising and financing food services

In general in developing countries practical steps to set up feeding services especially for small to medium size enterprises are few and far between and, as yet, inadequate to meet the needs of workers. The following paragraphs summarize the types of actions taken and the experiences of employers', workers' and governmental organisations.

In Latin America, many existing feeding services are set up by the joint action of employers. There are several cases where employers contribute a certain percentage of their payroll, say 1.5 per cent, to a fund which provides for the operation of public restaurants. Public restaurants are also set up in industrial areas by non-profit organisations and may be open to the public and workers in the area. "People's restaurants" set up by municipalities for workers (and sometimes their families) in a town or city are examples of this type of service. Other forms of organisation include employer co-operation in the establishment of central kitchens which cook and deliver meals for workers in member firms.

Sharing of canteens is another practical arrangement in some industrial cities. This is an arrangement whereby a small enterprise arranges with a nearby café(s) to provide meals for the workers at a special price in return for a guaranteed number of diners and a subsidy from the enterprise.

Inter-enterprise restaurants or canteens are in theory a good solution for small firms especially in new industrial zones. Some newly established industrial estates have canteens or restaurants in which the workers of all firms on the estate can eat. The rent paid by each enterprise to the estate management includes a proportion of the cost of running the canteen. The meal charges may cover the cost of food or the cost plus other factors.

Contracting out the catering is also an option taken in many countries. Firms sometimes contract out the catering but provide a company cafeteria or dining hall to which meals are delivered by a central kitchen, a contract caterer or a restaurant. This is useful for firms with premises large enough for a cafeteria but not for a kitchen, and it is one way of reducing capital outlay. There are cases of groups of enterprises employing less than 100 workers forming associations or co-operatives to jointly set up and operate feeding service centres to supply cooked meals to the enterprises from a central kitchen. The establishment of such "feeding service centres" is sometimes encouraged through subsidies and low interest loans.

A mess room where workers can eat food brought from home is another approach adopted by some enterprises. The capital outlay is much less than for a canteen: it takes up less land, it requires less staffing and the cost of services (utilities) and maintenance is less. Where some of the workers are not able to bring food from home, it may be feasible to arrange for an approved hawker or other food vendor to sell food (inside or just outside the factory gates) which can be eaten in the mess room. Sometimes the mess room attendant can be allowed to make and sell snacks to the workers. Although a mess room must never be used as a workplace it can be used for training classes, recreational facilities, union meetings and so on and thus become a useful asset to the enterprise.

An inter-enterprise mess room can be provided jointly by two or more neighbouring enterprises for their workers. Where space is a problem the solution may be "separation by time" whereby the use of the mess room is reserved for certain groups (e.g. young workers, women workers, the workers of one enterprise, etc.) at different times.

The provision of a special bonus or restaurant coupons paid to all workers in firms without canteens is found in some countries. Some firms provide bonuses to workers on rotating shifts or overtime, while others provide them with a snack or sandwich. Others provide extra milk or a substantial snack for workers in jobs involving certain chemicals, such as lead, or heavy manual labour.

Worker-financed food services run by a co-operative of workers are found in some countries. The employer may provide the premises and utilities, and the co-operative may run the service. A co-operative may arrange for an outside contractor or vendor to deliver a fixed number of meals each day which are then served by the co-operative's employees or by the members themselves. Co-operatives usually depend on members for providing voluntary help in the form of clerical and accounting services.

Further, in some countries, there are traditional feeding arrangements such as "luncheon clubs" where a group of workers organise the joint purchase of their midday meals from a café or hawker, and employ a helper to serve the meals and keep their mess room clean and tidy. The employer usually provides the mess room, furnishings and services (utilities).

Another traditional arrangement found in some countries especially on mines or other projects where the workers live on site in bachelor status is for the employer to provide fuel for cooking, a simple kitchen and a mess room for groups of workers who purchase their food and employ a cook to prepare their meals.

Low-cost or fair-price shops which employers in large enterprises may be legally required to provide for their workers are sometimes handed over to workers or trade unions to be run on a co-operative basis. The employer bears the capital outlay and frequently provides a loan at low or no interest to enable the co-operative to buy goods at discount prices. Other forms of help include free utilities; help from the company's accounts department in book-keeping methods; help from the company's purchasing officers in buying goods, providing assistance in transporting goods, especially food, from the nearest wholesale market to the shops or facilitating training in the management of co-operative shops and other relevant subjects.

Role of social welfare funds in financing industrial feeding services

In some countries where workers are entitled to a share of industrial profits, a percentage of the profits that legally belongs to the workers is paid into a special fund for financing the establishment of various community social services for the benefit of workers and their families in industrial districts. In some others, the law obliges employers in certain industries to set up and finance industry-wide welfare funds.

So far, such social welfare funds as do exist have rarely been used for financing workers' feeding schemes because those concerned with their management, whether governments, employers' or workers' representatives, have felt that other needs, such as housing, health services and cultural and recreational activities are more urgent. Nevertheless there may well be a case for contributions from such social welfare funds towards the setting up of simple, small-scale services. Indeed this has been provided for in the Welfare Facilities Recommendation, 1956 (Section 26) which states that "... in the absence of other legal obligations concerning welfare facilities, such facilities may be financed through welfare funds maintained by contributions fixed by the competent authorities and administered by committees with equal representation of employers and workers".

A tabular presentation of different ways of providing and financing food services is given overleaf.

A tabular presentation of different forms of providing and financing food services.

I. CANTEENS

	Total Cost	Food Cost	Land	Buil-ding	Canteen Staff	Equip-ment	Servi-ces	Mainte-nance	Remarks
Employer	X								a) The employer bears the total cost
Employer			X	X	X	X	X	X	b) The employer provides land, buildings, equipment, services* and staff. The workers pay the cost of the food or, in some cases, the cost of food plus a fixed percentage, e.g. 10% to cover the transport of foodstuffs, etc.
Workers		X							
Employer			X	X		X	X	X	c) The employer provides land, buildings, equipment, services* but a workers' co-operative or trade union runs the canteen and workers pay cost of food and canteen staff.
Workers		X			X				
Employer		X	X	X	X	X	X	X	d) An employer-financed social security group runs canteens in individual factories, industrial estates and industrial areas of cities. Such canteens may be open to all workers and their families in an area. The workers pay subsidised cost of food.
Workers		X							
Employer			X						e) The employer provides land only. A trade union or workers' co-operative runs the service on a non-profit-making basis.
Workers		X		X	X	X	X	X	
Employer			←------	------	------	------	------	------→	f) A social-security fund financed by employers and workers runs feeding programmes for workers.
Workers		X	←------	------	------	------	------	------→	

* Services (utilities) include water, electricity, and gas; sewerage and garbage removal.

IV. HAWKERS

	Total Cost	Food Cost	Land	Buil-ding	Canteen Staff	Equip-ment	Servi-ces	Mainte-nance	Remarks
Employer			X				X		a) The employer may provide a hawker with free services in return for operating a stall at certain hours inside the factory gates. Hawker's prices are to be no higher or even less than in open market. Joint employer-worker committees check market prices regularly.
Workers		X							
Employer			X				X		b) A hawker may pay rent for a stall inside the gates and for services provided by the enterprise. See a) above for pricing.
Workers		X							
Employer		X							c) An employer may arrange, in co-operation with the workers, for a hawker, contractor or local café to deliver cooked meals to the enterprise daily. They may be supplied free or at subsidised prices to the workers.
Workers		X							
Employer			X	X		X	X	X	d) Workers co-operatives or "lunch clubs" may order meals from a hawker, contractor or café, to be delivered to the establishment. The employer may provide a mess room or dining room and the workers pay food suppliers (and sometimes the wages of an attendant to prepare the room and wash dishes, etc.)
Workers		X			X				

III. MESS ROOMS

	Total Cost	Food Cost	Land	Buil-ding	Canteen Staff	Equip-ment	Servi-ces	Mainte-nance	Remarks
Employer			X	X	X	X	X	X	a) The employer provides land, building, equipment, services, furniture and attendant(s).
Workers		X							
Employer			X		X	X	X	X	b) The employer provides land and materials for building and equipping a simple mess room. The workers form a self-help group to build and equip the mess room. The employer pays for attendant(s), services and maintenance of the mess room.
Workers		X		X					
Employer			X	X	X	X	X	X	c) The employer provides land, building, equipment, services, furniture. The attendant or steward is given a concession to cook snacks and sell them to Workers. The type of snack and the prices charged are agreed upon by employer and workers.
Workers		X							
Employer			X	X	X	X	X	X	d) Where there is a canteen, a section of it may be used as a mess room by those who bring food from home (or have a meal delivered to their workplace).
Workers		X							

II. INTER-ENTERPRISE CANTEENS

	Total Cost	Food Cost	Land	Buil-ding	Canteen Staff	Equip-ment	Servi-ces	Mainte-nance	Remarks
Employer		?	<------Shared with co-tenants------>						a) On an industrial estate, tenants pay rent which will cover the running costs of the canteen - the workers may have to pay full price of meals unless the employer subsidises the cost.
Workers		X							
Employer		?	<------Shared------>						b) Two or more enterprises set up a joint canteen for their workers. A joint management committee will decide how meals are to be subsidised.
Workers		X							
Employer		?	<------Shared------>						c) A canteen set up for workers in two or more small enterprises within the same industry. Same as (b) above.
Workers		X							
Employer		?	<------Municipality or non-profit-making body----->						d) A municipality or other governmental body or a non-profit-making body may set up a canteen, sometimes called a People's Restaurant, which provides full meals to all workers, and often their families, in an industrial district. The meals are either charged at cost or subsidised by employers and/or government.
Workers		X							
Employer			X	X		X	<---Sometimes--->		e) A trade union may set up a canteen for members working in a certain district. Employers sometimes provide premises or a low-interest loan to buy or rent them. Employers may provide know-how and financial guarantees to enable bulk purchases of food stuffs at advantageous rates.
Workers		X			X		<---Sometimes--->		
Employer		(Individual arrangements between large and smaller enterprises)							f) A large enterprise may allow workers from neighbouring small firms to eat in its canteen or may supply cooked meals. This arrangement is more common in centralised economies or where the small firms do work on contracts from the larger firm.
Workers		X							

V. RESTAURANT VOUCHERS, TICKETS OR COUPONS

	Total Cost	Food Cost	Land	Buil-ding	Canteen Staff	Equip-ment	Servi-ces	Mainte-nance	Remarks
Employer	X								a) The employer may purchase "restaurant vouchers, tickets or coupons" from a central organisation. These are issued to workers who can exchange them in part or whole payment for a meal in a restaurant or café of their choice. There may be tax concessions for firms who use such vouchers, tickets or coupons.
Workers									
Employer		X							b) Where there is no such system, an employer may make an arrangement with a local restaurant(s) for workers to be served an inexpensive meal in return for a subsidy.
Workers		X							

Chapter 10

Physical facilities

Location and site

When the decision to set up a canteen or industrial feeding service is made, a suitable location for the cafeteria and kitchen will have to be selected. Kiosks, trolley services, satellite cafeterias and vending machines will, in general, be located close to the workplaces.

In some countries plans for a new canteen are required to be submitted to the Factory or Labour Inspectorate. The Inspectorate should be informed of the first meeting of the planning committee and invited to be present, after which it should be kept in touch with subsequent developments.

An existing building may be converted into a canteen or a new building constructed, but in either case there are a number of matters which should be considered before work begins. They include the following:

- Ease of access for workers. The distance from the workplace to the canteen should, ideally, be no more than five minutes' walk. If this is not the case, extra time will have to be added to the mealbreak. The access way should be under cover if possible. In monsoon and other wet climates, the walkways to the canteen should have overhead protection against rain, and the floor surfaces should be properly drained or raised, to prevent water lying in pools on the walkway. In hot dry climates they should provide protection against the sun and dust. Where there are frequent dust and sand storms the walkways should if possible be planned so that they are protected from the prevailing winds.

- Ease of access for food suppliers and garbage collectors. It should be possible for vans, carts or lorries to draw up alongside the kitchen stores delivery bay or the refuse collection bay. The bays should provide some shelter against rain or sun. In places where local deliveries of food may be made by handcart or bullock cart the planners should note this fact, and the height of unloading ramps adjusted accordingly.

- Ease of access for emergency services. Fire fighting equipment and ambulances must have ready access to the canteen building on both sides.

- Doors and exits. Canteen buildings and kitchens should have at least two doors 0.75m (30 in.) wide, which open outwards and lead to the exterior and are on different walls and are never bolted or locked during working hours. One door should be in the cafeteria, another in the kitchen. In large kitchens a second door from the supervisor's office to the outside is desirable (see page 128). Each door must have a clear sign (lit at night) EMERGENCY EXIT over it. Emergency Exits on upper floors must lead directly to a stair or gradual ramp. There must be a horizontal space (landing) equal at least to the width of the door at the top of the stairs. There should be an adequate number of the appropriate type of fire extinguishers and fire blankets in kitchens. Staff must be trained in their use and in orderly evacuation of premises through Emergency Exits. Untrained staff may tend to panic. Local Fire Brigades should be consulted at the planning stage about local fire regulations and for advice about Emergency Exits, fire extinguishers and training.

Water supply. It has been said that the single most important contribution to the health and efficiency of workers is the supply of pure, safe drinking water, and it is laid down in the laws of many countries that such a water supply must be provided.

Clean water must be available for drinking and the cooking of foods that have a short cooking time; and water is also necessary for the preparation of vegetables, dish-washing, floor cleaning, etc. Where an adequate mains supply of drinking water is not provided by the local authorities the establishment will have to provide its own drinking water supply - perhaps by sinking a tube well and installing a pump. The well water should be filtered and, if necessary, boiled or chlorinated or treated in some other appropriate manner before being used. Water-borne infections can cost many work-hours and affect the health of the whole community. The well must be strictly controlled and no unauthorised person allowed to use it.

The national and local public health authorities will be able to provide guidance on the filtering and bacteriological control of water. In addition, soft-drinks manufacturers, especially those with a franchise from a major company, often have considerable expertise in this subject and may provide information on sources of filtration equipment.

Potable water can also be bought in by road tanker from government or private sources. Whatever its source, the water must be stored in covered hygienic containers.

In the cafeteria, as at the workplace, if the drinking water is not piped to a tap, fountain or water cooler, it should be provided in metal containers (not in earthware containers as these have a tendency to encourage the growth of bacteria). The metal container should be fitted with a tap and be raised above ground level, for example on a bench or table. A supply of disposable cups should be provided or the workers may use their individual drinking mugs. The use of communal drinking cups must be prohibited.

The best arrangement, especially in the cafeteria, is for the drinking water to be supplied by a drinking fountain. In hot climates the water should be cooled.

Canteens on upper floors of buildings

Where space for a canteen is limited, the solution may be to build a canteen on top of an existing building. Several factors have to be considered before deciding to do this. These include:

The additional weight of kitchen equipment and of kitchen staff and customers, which may together be heavy enough to require extensive structural alterations.

Lifts. All food coming in to the canteen and all garbage going out will have to be carried in a lift and it may be necessary to install a service lift;

There may be additional use of the passenger lifts if customers are coming from other buildings to the canteen.

If handicapped workers are employed, care should be taken that lifts are wide enough for wheelchairs.

Fire precautions. The fire escape route must be adequate for the maximum number of people in the building, including the cafeteria, at any one time.

Two-storey canteen buildings

Where space at ground level is limited it may be feasible to build a small two-storey canteen with the kitchen and cafeteria on different floors. Three points to consider are:

(a) if the kitchen is on the upper floor, all stores going in and garbage coming out will have to be carried by hand or in a lift;

(b) if the cafeteria is on the upper floor, all employees will have to climb a flight of stairs unless there is a passenger lift. Access for handicapped workers should be provided;

(c) all food when cooked will have to go up or down from the kitchen to the cafeteria and a food service lift will be necessary.

The provision of utilities

In existing factories the ease with which drains, sewers, water, steam, piped gas and electricity can be connected to the canteen building will affect the choice of site. An engineer can advise on this.

If oil, bottled gas or solid fuel is to be used for cooking the extra space needed for fuel storage should be taken into account at the planning stage.

Climatic considerations

The positioning of the canteen building should take climatic factors into account. The best positioning of windows differs according to the latitude. In equatorial regions it is usual to put windows on north and south walls whereas further away from the equator it may be preferable to put them on the west and east walls.

In addition, in hot climates it is desirable to provide shading outside a building, by means of verandas, balconies, open wall construction, wooden screens, bamboo "chicks", hedges or leafy trees, for example, and to design the building so that it catches as many cooling breezes as possible. In cold climates and windy climates doorways and entrances may need to be protected from wind, rain and snow by porches, walls and double doors.

Low buildings surrounded by high ones

Where it is planned to build or convert a one or two storey building surrounded by blocks of offices, factories or warehouses that are several storeys high, the following should be considered at the planning stage.

Smells and odours: would the smell of food being fried pervade the offices in the vicinity? An appropriate system of ventilation may be necessary.

Light: would the higher warehouses block out most of the daylight from the canteen? Would it therefore be necessary to use artificial light all day?

Wind and natural ventilation: would either of the taller blocks affect the wind pattern causing icy blasts of cold air necessitating extra heating or insulation in cold climates, or blocking cooling breezes and making it necessary to use air-conditioning in hot climates?

Size of canteen building

In some countries government regulations lay down the minimum size for a cafeteria and kitchen and may stipulate that it should be able to feed a certain number (usually about one-third of the total workforce) simultaneously.

The space allowances given below should be regarded as a minimum, especially if the site is irregular in shape or the building has low ceilings. In some national regulations a minimum ceiling height is stipulated.

Cafeteria*

25 persons or less: 200 sq.ft; or 18.5 m^2

26 to 74 persons: 200 sq.ft. or 18.5 m^2 plus 7 sq.ft. or 0.65 m^2 for each person above 25;

75 to 149 persons: 550 sq.ft. or 50 m^2 plus 6 sq.ft. or 0.55 m^2 for each person above 74;

150 to 499 persons: 1,000 sq.ft. or 92 m^2 plus 5 sq.ft or 0.50 m^2 for each person above 149.

Secondary uses of canteens

A canteen can serve other purposes if it is so planned, e.g. it can be used for such activities as training classes, joint employer/worker meetings, union meetings and safety committees.

Tables and seats should not be fixed to the floor. Easily moved chairs and tables (preferably stacking) are the best furniture when it is planned to use the cafeteria for other purposes than eating.

Kitchen floors

Local conditions will determine the construction of floor to a certain extent, but floors should always be made of easily cleaned impermeable non-slip materials and laid with a slight slope. A kitchen floor has to withstand hard wear, water, grease, oil, acids, alkalis and cleansing agents and variations in temperature. If possible, too, it should be neither excessively hard or noisy. Materials in common use include quarry tiles, mosaic tiles laid in impervious acid-resisting cement. Terrazzo and granolithic screeds incorporating carborundum are also common.

---

* ILO Model Code of Safety Regulations for Industrial Establishments. Ref. 217.52.

To facilitate washing, all kitchen and store room floors should be laid so as to fall towards drainage outlets in the form of channels or gulleys (1 in 120 slope) which are covered by removable gratings. Where there are central islands of cooking equipment it is common to have drainage channels around the equipment to make cleaning and floor washing easier. All junctions between floors and walls should be curved, preferably using the flooring material. There must be no sudden changes of floor level in the kitchen area or passages leading to it.

Kitchen walls

The lower part of the walls up to about 2 metres or 6 feet must be water-resistant and easily washed. Glazed tiles are excellent for this purpose. A hard gloss paint on a dense smooth surface is satisfactory although it is liable to crack over large areas. Many kitchens are tiled up to the ceiling, and while this simplifies cleaning it can be noisy and lead to condensation. A light-coloured impervious washable paint is an alternative. Splashbacks behind stoves, washbasins and grills may be of glazed tiles, enamelled steel, stainless steel, glass or one of the plastics specially formulated for the purpose.

Partitions

Partitions are used to separate working areas in a kitchen and, (if not weight-bearing) can be of laminated plastic, stainless steel or anodized aluminium or sealed hardboard faced with plastic film mounted on light-weight frames. Weight-bearing partitions must be made strong enough for the load they carry. Partitions must be easily cleaned and dismantled to permit changes of the kitchen layout.

Ceilings

Routine cleaning of the kitchen ceiling is important thus it is wise to design the kitchen so that the ceiling can be easily cleaned.

Windows and ventilation

Correct ventilation in the kitchen is very important - it improves working conditions, increases efficiency and provides a healthy atmosphere.

In tropical and other hot countries the art of ventilation requires considerable skill and a specialist should be consulted if available, but a combination of local know-how plus basic engineering skill may be successful.

Special points to be watched include:

The balance of air pressure between the cafeteria and the kitchen. Air should be drawn from the cafeteria into the kitchen, not the reverse. In practice, this should not be a problem if the extraction ventilation in the kitchen is working properly.

The height of ceiling fans: at the wrong height fans may force warm air back over the people eating in the cafeteria. Inlet air blowers at floor level or "floor fans" may be more effective as a means of cooling.

A windowless canteen or kitchen is to be avoided. As mentioned earlier the correct positioning of the windows according to the climate is important. The windows should also be placed so that maximum control of the ventilation is possible. In the tropics, two rows of windows, one rather low with tilting panes permit easy ventilation control in the cafeteria. In the kitchen, because of the placing of sinks or cooking equipment against the walls, windows are not very useful as a means of ventilation. Extraction ventilation (to be discussed later) is more suitable.

In areas where the prevailing winds change seasonally to a marked degree it is sometimes possible to make best use of whatever cooling breezes there are by changing the hinging of window shutters. At the construction stage this can be provided for at little cost.

In desert areas there has recently been an upsurge of interest in the traditional "wind towers" and these have been installed in modern buildings as an economical means of assisting cooling and ventilation.

In cold climates or continental type climates with hot summers and cold winters, some form of heating will be needed for the cafeteria, and for parts of the kitchen, such as the vegetable preparation areas and storerooms when staff are working in them. Attention should be paid to the siting of outside doors so that they are not exposed to the full force of the prevailing winds and the use of porches or verandas to protect the entrances: this can reduce heating costs and improve comfort in the cafeteria.

Air-conditioning is the best means of controlling temperature and humidity but requires electricity and a suitably designed (or modified) building. Unless all the buildings in an establishment are air-conditioned the cafeteria need not be air-conditioned. There should not be too great a contrast between the temperatures of the working areas and the cafeteria.

Where energy costs are high or electric power sources unreliable, a building designed according to the traditional, time-tested standards of a hot climate is to be preferred to one designed for air-conditioning.

The more traditional methods of keeping a building cool involve the use of space. A veranda is usually 1.5 metres (5 ft.) deep, with porches and double doors. It is very important to consider their use at the planning stage so that enough space can be allocated for them. The additional building costs of a veranda will not be high if it is in the plans from the beginning and this will, of course, save fuel.

Exhaust ventilation over cooking equipment is essential and must be properly installed. Reputable manufacturers of cooking equipment will state the exhaust ventilation requirements of their equipment. Exhaust ducts should be properly installed when the canteen is built and the electricity supply should be adequate to permit further extractor fans to be added simply and cheaply at a later date. Simple fuctional hoods or canopies conforming to good national practice should be installed over the equipment in the kitchen and over the serving counter in the cafeteria if there is a hot cupboard or grill or bain-marie giving off heat. The height of extractor hoods should be such that there is no risk for workers: hoods should be fitted 1.5 metres (5 ft.) above the equipment and at least 2 metres above (6 ft. 3 inches) if workers pass under them.

Hoods and canopies need to have a filter to collect grease and absorb cooking smells, which is easily accessible for cleaning and replacement. In order that canopies and hoods do not obstruct the natural light they can be made of translucent reinforced glass or fibreglass, or otherwise they can be made from stainless steel, aluminium or galvanised steel.

An important consideration is that the ducting from the hood or canopy in which there is an extractor fan must be constructed so that it vents to atmosphere outside of the building. This vent should be at a location and height such that the exhausted kitchen air does not create a nuisance for workers in other parts of the building. Extractor fans must be sufficiently powerful and in a large kitchen it is wise to have a spare fan in reserve.

The traditional method of roof insulation in tropical countries is thatching or earth on which grass is grown, or a flat roof on which water is sprinkled. These methods have drawbacks in industry - thatch is a fire hazard and in a drought the grass on a roof "garden" could also catch fire. Sprinkling is not very practical. Double roofs are an answer but they need to be carefully designed to avoid risk of fire and should conform to the highest standards of fire protection. Heavy-construction roofs supported by heavy-construction walls provide good roof insulation and are to be preferred to thin flimsy roofs. Thought can be given to the possibility of having a small canteen built in the traditional manner by local contractors but to insist on rigourous inspection to ensure that the building meets modern fire, safety and hygiene standards.

Fly-screening

When building a canteen in regions where fly-screening is necessary, it is desirable to fix the screens a little distance from the outside of the window and to have the windows opening inwards. Where there is a balcony or veranda, it should be screened, rather than the windows. Wire mesh should be held tightly in a metal or timber frame fitted to the opening without any gaps.

External doors must also have separate screen doors opening outwards. It is especially important to fit screen doors on kitchen and storeroom entrances and on the storage areas for garbage cans. A self-closing device, which can be quite simple, should be fixed on screen doors and should be checked routinely for efficiency.

When large numbers of people enter the canteen at certain times of day it is very difficult to keep out flying insects unless an "air curtain" can be provided by one or more fans fixed so as to blow air down over the entrance. Bead curtains play a role in keeping some flying insects out of a room but cannot be regarded as a substitute for proper fly-screening. To be effective fly-screening must be kept in good condition - one very small hole can allow sufficient insects through to start a breeding colony - it is therefore prudent to make provision for maintenance of the screening and to have reserve mesh in stock.

Lighting

The cafeteria, service counter and the kitchens each have their own special lighting requirements. Guidelines are available based on long experience and investigations by ergonomists and others. These should be consulted when planning the lighting of the canteen. Extra electric points for lighting should be built into the canteen so that modifications can be made after the canteen has been in service.

In the kitchen good lighting is essential - for hygiene, for accurate working, and to prevent accidents. Where it is readily available fluorescent lighting, while more expensive to install, is usually cheap to run and maintain and unlikely to produce much glare. Additional light fittings may be needed in or around canopies and hoods. Since the accumulation of dirt on electric fittings reduces the amount of light given out, it is important to arrange for regular cleaning at times when the kitchen is not busy. There are several "colours" of fluorescent tube available and some of these are better for inspection work, others better for lighting serving areas, and others more suitable for general purposes. Information can be obtained from suppliers or manufacturers.

Lighting over the serving counters should be good and here if fluorescent lighting is used it is very important to use the correct colour tube, as the wrong colour may give food a most peculiar appearance and give rise to complaints from customers. If the lighting at the serving counters is insufficient, the service will slow down and standards of hygiene may deteriorate.

In the cafeteria itself the lighting need not be as bright as in the kitchen and serving areas, as a lower level of illumination is more restful. In a large canteen bright areas and less bright areas can be achieved by adjusting the wattage of the light bulbs or tubes.

Agreement should be reached with the maintenance department as to whether it or the canteen staff are responsible for replacing old light bulbs and cleaning fittings in the canteen. If the canteen staff are to be responsible then they must be provided with a suitable step-ladder and the canteen supervisor must allocate this duty to certain of the staff.

As an example of recommended lighting standards the International Electrotechnical Commission Standard Code for Interior Lighting states that food stores should be provided with approximately 150 lux, while general working illumination, measured on a working surface, should be around 500 lux.

Sanitary facilities

Toilets (WCs) and washing facilities should be provided at the canteen or close by, and considered at the planning stage of the canteen. As good personal hygiene among kitchen staff is very important, it is essential that fully adequate washing and toilet facilities be available to them.

Matters to consider at the planning stage include:

- the national laws regarding sanitary facilities in factories and canteens where they exist;
- special provision for women workers , whether or not any are employed at present;
- provision of a shower for the canteen staff, which will be appreciated in areas where washing facilities in the workers' homes are limited;
- local customs such as the use of fingers for eating rather than using cutlery or chopsticks, rendering installation of wash-basins at the canteen entrance necessary;

- construction of a toilet cubicle for use by a handicapped person: this usually involves fitting sliding doors wide enough to admit a wheelchair,provision of hand holds, a cuvette type urinal, a raised toilet seat and a wash-basin cantilevered out from the wall.

In the kitchen itself at least one wash-basin should be provided for the use of staff. Large establishments will require additional basins near the food preparation areas. These basins need not be large - and indeed should be small enough to prevent their being used for washing dishes or pans. Alongside the basin there should be hand-drying facilities which must not be towels for communal use, but hot air dryers, or roller towels providing an unused portion of towel for each user, or paper towels or individual and clearly marked hand towels on separate hooks for each member of the staff.

Separate facilities for men and women

Where the regulations require a separate eating facility for women workers, or it is customary for men and women to eat separately, a section of the cafeteria can be screened off and used as the "ladies corner". In this case some form of waitress service or a separate small serving counter is necessary to eliminate the need for the women to stand in line at the main counter to collect their food.

Rest areas

National regulations may require the provision of such rest areas, especially for women staff, and even in their absence their provision is advisable for kitchen staff, who have to work in very tiring conditions in a hot, humid atmosphere. Where the staff are not accustomed to wearing shoes indoors in their own homes they may appreciate a small area, perhaps on the veranda or a corner of the rest area which is covered with a mat, on which they can relax after removing their shoes.

Canteen advisory services

An establishment wishing to provide a canteen for its staff may seek the services of an independent advisory service for planning guidance. Where such services do not exist there may be a government factory canteen advisory service, or a school meal service may be willing to give help on choice and planning of equipment to a small firm.

Many countries have hotel and catering management colleges and the staff of these institutions will have the expertise and knowledge to help and advise on industrial feeding services; but may be so busy with their own work in developing tourism or hotel management that they may not have time to help. Professional associations of restaurateurs, architects, engineers, doctors, environmental hygienists may be able to help. Manufacturers of catering equipment sometimes have their own consultancy services. But it is likely that the small to medium-sized enterprise which wants to set up a small industrial feeding service for its staff may do much of its own planning, limited by a budget and the space available for the canteen building. In the section on planning layout, some examples, in diagram form, have been given of canteen layouts which have been used elsewhere. These would have to be modified to take into account the type of food to be served, the amount of space available and the amount of equipmennt that can be bought. Just as on the factory shop-floor sensible arrangement of machines, gangways, storerooms makes for more efficient working, in the canteen kitchen these same factors have to be taken into account.

Cafeteria counters

The serving counters can range from a simple hatch in the wall from the kitchen, to elaborate counters complete with micro-wave ovens, deep fryers and grills in which complete meals are cooked to order, while the customer waits.

For most canteens, a serving counter with a limited amount of equipment will have to suffice. The equipment chosen will depend on the menu and on the climate - in a cold climate plate warmers are necessary and may either be in the form of hot cupboards or self-contained electrically heated mobile plate dispensers. In a hot climate the main concern may be to keep the counter and the area surrounding it as cool as possible. Hot drinks such as tea and coffee may be dispensed out of large domestic type pots or sophisticated 'pressure boiler' or 'expansion boiler' type cafe sets. Many variations of these basic layouts are possible and they can be used to provide discreet separation for workers of different ethnic groups or religions.

Colour

Commercial restaurateurs have long realised the value of pleasing colour schemes in attracting and retaining customers and, in industrial catering, when a wall or ceiling has to be painted, the choice of an attractive colour involves no additional expense. Since colours have emotional and religious significance in some regions of the world, the final decision on what colours to use should be discussed with workers' representatives. In general, blue and green tones give a feeling of coolness while red tones give a feeling of warmth. Light tones are advisable in kitchens and cafeterias. Washable paints should be used. Tables and chairs are often available in colour at little or no extra cost; or old or second-hand chairs and tables can be painted.

Curtains

Curtains over windows are necessary when the cafeteria is reserved for women at certain times or in a separate dining-room for women. If curtain rails are installed when the cafeteria is being built or renovated it is then a simple matter to hang up curtains made from cheap locally available material. The chief objection to curtains occurs when smoking is permitted in the canteen. Curtains made from flame-retarding material should then be used.

Bead curtains are very popular in some tropical countries and have the merit of providing privacy without blocking out air movement. If chosen in attractive colours they enhance the appearance of the room.

Pictures. Often there will be amateur artists on the staff who, given a little encouragement, will enter their paintings, drawings, collages, etc. into a company-sponsored exhibition. The exhibits could then be hung for a month or so on the cafeteria walls, all that is required being some simple means of hanging the pictures. Paintings or drawings by children at a local school or those entered in a competition for employees' children could also be exhibited.

Office for canteen manager or supervisor

A small office adjacent to the kitchen or partitioned off within the kitchen is necessary for the supervisor. Apart from desks, chairs, adding machines, coat rack, fan or heater, it is desirable for the office to have a window looking out into the kitchen so that the supervisor can keep a constant check on the hygiene, tidiness and safe working practices of the staff. The office will require a telephone, with an extension into the kitchen.

Building-in safety

Just as in the factory, the way in which the shopfloor is laid out can affect safety in the canteen too. Falls can be reduced by such measures as:

- choosing anti-slip surfaces for the kitchen and cafeteria floors;
- ensuring that there are no sudden changes of level in the kitchen floor and that the floor is maintained in good condition. Broken and cracked floors can lead to falls.
- allowing sufficiently wide gangways and passages between equipment.
- ensuring there are no blind corners round which persons carrying trays or pushing trolleys could collide;
- ensuring that doors between cafeteria and kitchen are either made of translucent plastic or have windows or portholes of wired glass or perspex;
- ensuring that the size of equipment in the kitchen is suited to the dimensions of the average worker in the canteen. Strain and stresses caused by having to bend too low over sinks or reach too high into ovens can lead to accidents;
- providing a mop, bucket, long-handled dustpan and brush in a convenient location so that spills and breakages can easily be mopped up without loss of time.

The canteen staff should be included in any safety competitions or promotional activities held within the establishment or group of establishments and in connection with any national safety week. They should be eligible for awards for safe working, in the same way as other workers.

Chapter 11

Equipment and maintenance

Introduction

It is important that cooking appliances should be of a suitable size for the cooks using them. Imported equipment or locally-made copies of imported equipment may not be suitable in this respect.

Equipment for the preparation of food before cooking (or of raw dishes) is designed to do one or more of the following actions: washing, peeling, slicing, chipping, mixing, chopping, grinding, sieving, mincing. Examples of hand-operated as well as electrically powered large-scale equipment are given below. For small canteens the hand-operated equipment may be suitable for several tasks, depending upon the menus chosen.

The principal methods of cooking are: boiling, frying, steaming, braising, grilling, baking, roasting. Some examples of the necessary equipment are shown below.

Process	Traditional	Industrialised
Boiling	Pans, cauldrons, kettles Boilers with integral coal or wood fire	Boiling pans, tilting kettles, bratt pans, etc.
Frying	Woks, shallow frypans UNSAFE (Deep fry pans over open fire, electric or gas rings)	Bratt pans (shallow fry) Thermostatically controlled pans with safety cut-out
Steaming	Sieves over boiling water	Bratt pan Steam cooker
Braising	Cast-iron pans	Bratt pans
Grilling	Vertical spits Racks over charcoal	Gas, electric grills
Baking	Wrapping food in leaves Coating in clay and cooking on hot stones or embers	Gas, electric, solid-fuel baking ovens
Roasting	On a spit in front of or over a fire	Ovens

Examples of industrial cooking equipment

Boiling tables. These may be gas or electric, placed on top of ovens or counters or freestanding. They may comprise a mixture of gas and electric rings, or a solid top. A griddle for cooking eggs, scones, sausages, etc. can be built in. The maximum capacity of pans for ease of handling is 14 litres and should be less for staff of small stature.

Boiling pans or steam kettles. These are jacketed containers heated directly by steam (from a central supply) or heated internally by gas or electricity. They are easily controlled from a gentle simmer to vigorous boiling. Water is supplied by direct connection or swivel faucet (tap) and emptied by stopcocks or a tilting mechanism. Sizes vary from 45 to 137 litres. Boiling pans are suited for very large operations.

Tilting kettles. Similar to boiling pans, more flexible in use and easier to clean. They consist of double pans heated by steam, ranging in capacity from 13.6 to 400 litres. They cook very quickly without risk of burning. A hot or cold water supply is incorporated. Tilting is effected by lever or hand wheel. It is considered that two smaller tilting kettles do more than twice the work of a larger size because of the flexibility they provide in menu planning. Plenty of space is needed in front of a tilting kettle for safe working.

Deep-fat fryers. Fryers must be well constructed and should conform to the most stringent safety standards. For gas-heated fryers the recommended ventilation system must be installed.

Fryers are made in a wide range of size but since cooking time is short a small model is often adequate. A fryer should have a visible external thermometer showing the temperature of the fat, a thermostat to keep the temperature constant and a cut-out control in case the thermostat fails to work. There should be an easily accessible draining tap with incorporated filter for draining off the oil after frying. A close-fitting lid which in emergency can be put over the pan quickly if it catches fire, a small fire blanket and a small dry powder fire extinguisher should located close at hand. It must be possible to switch off the heating source (gas or electricity) without reaching over the pan. In a small canteen frying at the domestic level can be done with the traditional deep pans and baskets of the district but the necessary safety precautions must be observed. A purpose built fryer will produce better food in the hands of a less skilled cook than the "domestic" methods because of the thermostatic control of temperature.

Bratt pans. These heavy rectangular pans are mounted on trunnions to allow tilting. Heating is by gas or electricity and is thermostatically controlled. They are very versatile and can be used for braising, shallow frying (fat to a depth of 2 inches), griddling, poaching, stewing, boiling (water to a depth of 3 inches) and pot roasting.

Steamcookers or steam ovens. These useful ovens may operate at atmospheric pressure and handle large quantities of food (e.g. cooking potatoes, puddings) or at high pressures (0.5-1kg/cm^2 or 7-15 psi), so that food, especially vegetables, can be cooked in small quantities as required. Atmospheric pressure ovens can also, with suitable precautions, be connected to a central steam supply.

Forced convection ovens. Gas or electric. An incorporated fan forces hot air round the oven, ensuring and even temperature and faster cooking. Large ovens can have racks with removable shelves that can be rolled on or off trolleys, thus minimising handling.

Microwave ovens. Worth considering for use by small night shift.

Baking ovens. Only necessary where it is planned to bake all bread and cakes in the canteen. Otherwise a local baker may be able to supply all the baked goods necessary.

Grills and broilers. May be useful where there is a separate executive dining room or club where grilled foods are cooked to order. In some regions vertical spits may be useful. Salamanders grill the top of food and are useful in conjunction with microwave ovens.

Wire baskets. These can usually be made to measure by a local metal worker to fit various items of equipment such as pans, fryers or tilting kettles to enable small quantities of food to be cooked easily. Baskets can be rectangular, square, circular or quadrants of a circle as required.

Small kitchen utensils. Allowance for the purchase of small essential items such as knifes, strainers, spoons, etc. should be made by the planning committee. A shadowboard for utensils can be made by a carpenter; a lockable one for knives is desirable unless a knife block is used.

Potato peelers. These can be also used for root vegetables. In some areas potatoes and vegetables can be bought ready prepared. A double sink, preferably on heavy-duty wheels for easy moving, is needed for vegetable preparation. Before installing a peeler, check with the local authority about drainage, etc.

Slicers. Use for slicing meat, charcuterie and bread.

Chippers and vegetable cutters. May be a separate machine or attachments to the mixer.

Mincer. May be a separate machine or an attachment to the mixer.

Mixing machines. Electric - ranging in size from 5.6 to 60 litres. They may be table or floor mounted. Attachments for slicing, grating, grinding, mincing, shredding, chipping potatoes, grinding etc. are available. A second bowl is useful. A wheeled "dolly" in which to sit the mixing bowl saves time and energy. It may be locally made.

Cleaning equipment

Equipment for cleaning kitchens and cafeterias should be of appropriate size, made of solid materials and designed for the job it is to do. A locally made brush of dried grass may do the job well. A wide choice of purpose-designed brooms and brushes is available.

Electric floor polishers, vacuum cleaners and scrubbers, whilst desirable, have disadvantages - they take a minute or two to set up and are noisy and thus have to be supplemented in a cafeteria and kitchen by floor mops, long-handled dustpans, sponge mops, buckets with wringing devices. Divided buckets and buckets on wheels save time and labour. Walls can often be cleaned without the use of step-ladders or scaffolds if extension rods or poles (up to 6 metres or 20 feet in length) are used. Rubber-bladed squeegees for window and floor cleaning with blades ranging from (30 to 75 centimetres or 12 to 30 inches) in length can be obtained. It is important to select the right length of blade.

Gastro-norm and other standard sizes

At one time sizes used by kitchen equipment manufacturers varied considerably and it was difficult to interchange trays, baskets, and so on.

There is now however a widely used standard for containers known as Gastro-norm and many manufacturers specify that their ovens, steam coolers, bain-maries, etc. conform to Gastro-norm.

Tables and trolleys

These should be of impervious, easily washable, light but strong material. Stainless steel is very suitable. Whenever possible the trolley or table should be fitted with large heavy-duty wheels with brakes.

Trolleys vary in size but in general should not be larger than 0.6 x 0.9 x 0.9 m. or 2 x 3 x 3 feet. The quality of the wheels (castors) is important and they must be strong enough to withstand very heavy loads. Smaller trolleys but extending to 1.4 m, or 4'6" in height are used for lighter loads. Care should be taken that the trolleys are not too wide for any of the doors and passages, and that the height of the top shelf is the same as the counter-tops and preparation tables, so that food and equipment can be easily moved to and from the trolley.

Tables should not be more than 75 cm (2'6") wide, although there are exceptions. If a table is longer than 1.8 m (6 feet) it will need extra legs or supports.

The use of wheels, even as suggested above for mobile sinks for vegetable preparation and pre-cooking storage, is a great asset in a canteen. Trolleys can also be used for taking dirty dishes to the dishwasher and from there to the counter.

Trolley and table wheels must be fitted with brakes (or be lockable) and if such wheels are not available locally an attempt to obtain them from elsewhere should be made. A trolley or table cannot be used as a worktop if the wheels are not braked (locked).

Dishwashing

Manual dishwashing is only truly satisfactory when carried out by conscientious, keen-sighted and unhurried workers. While dishwashing machines have their faults, on the whole they work satisfactorily, day in and day out and even in the smallest kitchen prove their worth over the years. If their acquisition is not possible for financial reasons when a canteen is set up, it is wise to plan for the future by allocating space for one and ensuring that the necessary power (and perhaps gas), water and drainage outlets are built in.

The machines come in a variety of sizes and designs and it is possible to find a model which will fit into the most awkwardly shaped kitchen. Tabling, preferably in stainless steel or polished aluminium must be provided at both sides of the machine for dirty dishes and for clean dishes as they come out of the machine.

Some machines are almost entirely automatic - the dishes move in racks on a conveyor belt through a pre-wash, wash, pre-rinse and final sterilising or sanitising rinse at 77°C (170°F). In other models the dishes are first hand-sprayed or scraped, put into racks in a sink where mechanical brushes or mechanically agitated water cleans them, then into another sink and finally the racks are put into the machine for the sanitising rinse, after which they are removed and left to dry before being transferred to cupboards or the serving areas.

In other cases the dishes are hand-sprayed or scraped over a sink, then packed into racks which are slid by hand into a machine which automatically washes, rinses and sanitises them. Hot air drying is an optional feature for cold climates where plastic tableware is used. Trays of the appropriate size can also go into the dish-washer. Many serving dishes will go into a dishwasher but there will still be a need for a manual pot and pan washing section in a small to medium-sized kitchen. Pot-washing machines are only justified in very large establishments or central kitchens preparing food for a cook/freeze or cook/chill programme.

Dishwashing by hand. Not a pleasant job, dishwashing can be very tiring in hot, humid weather and only rarely is the dishwasher given the full recognition for the work he does. The work nevertheless plays an important part in preserving the health of the workers. Breakage of dishes is higher when dishwashing is done by hand.

Layout for dishwashing by hand. Three sinks (of which the first may be a small "half-sink") are needed.

(a) In sink 1 an overhead water spray flushes the plates, etc.;

(b) they are put into racks with long heat-resistant handles;

(c) the racks are put into sink 2 in water at 48°C (120°F) with detergent;

(d) the dishes are sprayed by a powerful overhead water jet or brushed by hand;

(e) the baskets are lifted into sink 3 into water at 77°C (170° F) or into water at 71°C (160°F) plus a sanitizing (sterilising) solution (NOTE Long-handled baskets are essential because the sanitizing water can scald a person very badly.);

(f) the baskets are lifted out of sink 3 and left to drain before stacking and storing.

Pan-washing by hand. At least two strong heavy-duty sinks large enough to hold the largest piece of equipment in regular use are required. The procedure is:

(a) the pans are soaked in sink 1;

(b) they are then flushed with an overhead water spray;

(c) sink 1 is filled with hot water and detergent and the pans are hand-washed;

(d) the pans are then put into baskets with long heat-resistant handles and immersed in hot water in sink 2;

(e) the water in sink 2 is heated (by steam or by gas or electric heater) to 77°C (170°F) for the final sanitising rinse. (NOTE This temperature is dangerous and can scald a person badly. Racks must be provided and used.)

Sanitising solutions are available which, when added to the water in sink 2, enable a lower temperature of water to be used.

Refrigeration and cold storage

In tropical and other warm climates a large volume of refrigeration space will be needed as the normal room temperatures favour the growth of bacteria and will be unsafe for food. The accompanying chart shows the safe temperatures for storage of different foods.

Refrigerators are used for short-term storage of food before preparation, between preparation and cooking and for safe keeping of cooked food for short periods such as overnight. Cooked food that is to be kept must be cooled rapidly before it is put in the refrigerator: in a hot climate this is difficult unless there is an air-conditioned room or fans.

In general the keeping of cooked food is to be avoided unless very good chilling facilities are available. Careful portioning and planning of menus will obviate the need to cook more than is needed on any one day.

The refrigerator should be large enough or larger than is necessary and should be well insulated. Some models for use in tropical climates have special insulation. A thermometer should be used.

Freezers are used for longer-term storage of food - they can be free-standing or walk-in. "Walk-in" freezers can be built to measure to fit into a kitchen. They can also be of modular construction and easily incresed in size as required. Thermometers must be fitted into the doors so that they can be read from the outside. There must also be a warning system - a bell which rings in the telephone exchange or a control room if the temperature rises because of a power cut or compressor failure. The doors of "walk-in" cold stores must be fully openable from the inside of the cold store.

Storerooms

The amount of space allocated to storerooms will depend upon the availability of local markets and upon whether it is possible to have deliveries of perishable foods daily, three times a week or weekly. The storeroom should be sited where goods are delivered. The storekeeper should have scales with which to check the weights of goods delivered.

Weighing scales

The storekeeper, or in a small canteen the supervisor, needs scales for checking the weight of the goods delivered. The cooks need scales for the preparation of food. There is much to be said for using simple balance scales with weights.

Garbage (rubbish) disposal and can washing

The least objectionable and most hygienic way of dealing with garbage or waste is to have:

- mechanical waste disposal units, plus plastic bags to hold all food waste which cannot be dealt with by the units; and
- mechanical waste compactors which compress all other waste into about one-third of its bulk which is then put into strong paper bags and sealed.

"Waste compactors and moisture extractors", which reduce food waste into an odourless pulp of about one-third the original volume, are also available.

The cost and/or local regulations may preclude the use of disposal units and compactors for many canteens and industrial feeding services. It is necessary to discuss the garbage collection and disposal arrangements at the planning stage.

Where plastic bags are cheap and plentiful it is much simpler and more hygienic to use them instead of garbage cans. The bags can be bought in bulk, and should be placed inside cans or in holders which ensure that they are above ground, and covered with a lid. When full, the bags should be tied tightly and stored above ground level in a fly-screened store easily accessible to the garbage collection service.

If there is no local collection the establishment will have to install an incinerator or make other arrangments for disposal of its garbage. Permission may have to be obtained to operate an incinerator and, even where this is not required, the local health inspector should be consulted as to the best place to site it. The composition of the garbage bags affects their suitability for burning in an incinerator.

While awaiting collection, the garbage cans or plastic bags should be kept in a shaded, fly-screened, vermin-proof area with a concrete floor, ridged to promote drainage and with a drain immediately outside. Food waste should be kept separate from other garbage and the supervisor should inspect the food waste regularly, and the reasons for an excessive amount of food waste should be investigated immediately.

A special space outside the kitchen or storeroom will be needed for storing clean crates, paper, glass jars and metal drums which are saleable or can be returned to the supplier.

Washing garbage cans

To eliminate the risk of infection, to prevent unpleasant odours and to keep the yard and kitchen entrance clean and hygienic, garbage cans must be washed after use. Automatic can-washers are needed only in very large establishments. For a smaller canteen the requirements are:

(a) a cold water supply from a standpipe or hose-reel attached to the nearest tap;

(b) a sloping pipe-rack on which to place the cans;

(c) a hard brush on a long handle to clean out the inside of cans properly (the brush must be shaped to get into crevices - an ordinary yard brush cannot do this). The type of brush that can be attached to a hosepipe such as is used for washing cars is useful. Remember the outside of the can must also be cleaned.

(d) the washing area must have a slope to a drain;

(e) there must be a grating over the drain to prevent scraps of food accidently getting into the drainage system and prevent vermin entering the yard through the drain;

(f) hot water or steam for the final clean in cool climates; in hot climates the cans can be dried out by the sun.

Food waste may be purchased by contractors for animal feeding, who will stipulate what items they are prepared to buy, and the garbage cans must then be marked with pictograms to illustrate what may or may not be put into them.

Seating and tables in cafeteria

The greatest flexibility in seating arrangements in a cafeteria is obtained using tables for two or four diners. These can be put together to seat six, eight, ten or twelve. In this way the tables can be used for meetings, training courses, etc, when meals are not being served.

Seating which is fixed to the floor, or is part of a table which is fitted to the floor, may simplify cleaning a little but has many disadvantages.

Tables may be arranged in straight lines or square tables may be arranged diagonally. This provides more space between chairs. If meals are served on trays, the size of trays chosen should be such that they fit easily onto the table. It is sometimes forgotten that the height of chairs and tables must be compatible - a high table and a low chair can be uncomfortable. The stature of the workforce should also be taken into account when choosing chairs and tables.

Maintenance of buildings and equipment

Maintenance of the canteen buildings and equipment usually becomes the responsibility of the enterprise's engineer (where a group of small establishments share a canteen, the responsibility for maintenance should be decided upon at an early stage). The maintenance of catering equipment requires the same skills as that of maintaining the machines on a production line, with the added imperative of strict hygiene and the difficulty of doing work when a meal is being prepared.

The engineers's co-operation is vital to the success of the canteen. The engineer may well be co-opted on to the canteen management committee, and should also be a member of the canteen planning committee. The engineer's schedule must allow for regular routine maintenance, and reports on this should be sent to the canteen management committee. All pressure equipment, steam boilers, etc. and lifts must be checked regularly in accordance with the national regulations, factory acts and/or the requirements of the insurance company.

Drains and gulleys must be checked regularly and cleaned or repaired as necessary. The kitchen walls, windows and ceilings must be washed or painted regularly. The kitchen staff may be responsible up to 1.5 m (5 ft.) above the floor but above that height the engineer's staff should be responsible (where long-handled equipment is available this rule can be modified). Painting of walls should be done at least as often as in the factory and offices.

The condition of walls and floors in all canteen buildings must be regularly checked by the canteen manager/supervisor for cracks and other damage, and any that is found should be reported in writing to the engineer and canteen management committee. Where there is a safety committee and/or a safety officer, the canteen building must be included in their regular tours of inspection.

When new equipment is installed the instruction booklet should be immediately photocopied three times. If a piece of locally manufactured equipment comes without written instructions the canteen supervisor/manager and the engineer should draw them up themselves in consultation with the manufacturer. Once photocopied, the original instruction booklet should then be given into the safekeeping of the purchasing officer or the accounts department.

One copy should be given to the maintenance engineer, one copy kept in a safe place in the supervisor's office, and the third copy kept in an accessible place in the kitchen. Where members of the staff are illiterate, pictograms should be used to illustrate how to operate and clean a machine correctly. Sometimes national safety promotion organisations offer inexpensive collections of pictograms for sale, and which may be adequate for this purpose. Otherwise a draughtsman or amateur artist on the staff can be approached for assistance in producing the pictograms.

It is not only the major, large items in a kitchen that need maintenance. Knives should be kept sharp - blunt tools are hazardous - and arrangements for their regular sharpening should be made. Trolley and table wheels need servicing from time to time if they are to run smoothly. Light fittings need to be regularly cleaned - otherwise much of their efficiency is lost.

In large towns with good maintenance services the regular servicing of equipment may be done by the manufacturer's own service staff. When selecting equipment it is prudent to check on the availability and reputation of the service provided.

If equipment is being imported with hard currency, the question of spare parts needs to be considered. It may be wise to order some spare parts with the machine if at a later date there is likely to be a long wait for permission to use hard currency. The purchase of locally made equipment has many advantages.

Chapter 12

Operating the canteen

Hygiene

In the previous chapter, emphasis was placed on the need to consider hygiene at the planning stage. It is indeed possible to maintain a fairly good standard of hygiene in poor surroundings but only the highly motivated, well-trained and reasonably well-educated person will be able to maintain high personal standards if the infrastructure is missing. It is unreasonable to expect that all staff of a canteen will be highly motivated and well-educated - it is assumed that they have been well-trained by the canteen supervisor - and everything should be done to help them to maintain a high standard of personal hygiene and food handling. The provision of wash basins in toilets and in the kitchen; hot-air hand dryers or some other hygienic form of hand drying; dishwashing machines or the appropriate sinks and sanitizing equipment; suitable waste disposal arrangements; all these help to ensure hygienic handling of food, dishes and cutlery.

Training in personal hygiene is important, and the most effective way of teaching is by way of films, videos or film-strips. These can be hired and shown to all the canteen staff. The showing can conveniently be combined with a film on safe working practices, for example, which would be as relevant to canteen staff as to workers on the shop floor. Regular annual film shows coupled with a lecture on food hygiene can have a cumulative effect.

Setting an example is an effective means of teaching high standards of hygiene - a supervisor who always washes his hands before handling food, and whose clothes are always clean sets a standard which is likely to be followed.

A problem that occasionally crops up is that one or more of the catering staff may be found to be a symptomless carrier of a pathogenic germ. Until they have been effectively treated, they should not be allowed to work in the kitchen, although they may be fit to do another job in the enterprise. Workers should not be dismissed for such a reason - not only because of the distress it could cause but because if one person is dismissed because of being a symptomless carrier it becomes very difficult to persuade the other staff to report any minor symptoms of illness, which may be the first signs of an infection that could be transmitted to the food and thus to the canteen customers. In some enterprises workers may report such symptoms directly to the nurse or welfare officer rather than to their supervisor.

Food should be handled as little as possible after cooking and then only with clean implements. Any cuts or sores on the hands or arms of kitchen staff should be covered with a waterproof dressing during working hours. In areas where it is customary for food to be touched or eaten with the hands certain taboos exist that ensure that in domestic situations this can be done relatively safely. In the canteen kitchen the quantity of food handled, the speed at which staff have to work, the temperatures, and the fact that it is almost impossible to eliminate a time lag between cooking and serving food, mean that more stringent rules of hygiene need to be enforced.

Correct methods of dishwashing, whether by hand or machine, are essential for the maintenance of a high standard of hygiene. They have already been discussed in some detail. If a dishwashing machine cannot be afforded, it is essential to provide at least the minimum of three sinks for dishwashing and two for washing pots. As relatively inexpensive items of kitchen equipment, sinks are important links in the chain of hygienic food handling. Wooden sinks are no longer considered desirable or hygienic. Where chinaware or pottery plates and dishes are in use, breakages are more frequent when their number is not sufficient to cover the main meal service, and when they have to be hand-washed and then carried to the cupboard or counter, instead of being put in baskets or racks into a trolley.

Eyesight has also been mentioned as a factor in hygiene. The health service should be asked to check the eyesight of kitchen staff before employment and by annual examination. The cost of buying a pair of prescription glasses for canteen staff is probably less than the cost of buying a hard helmet for a worker on a building site but just as important.

A simple check-list such as the following can be used to ascertain the general standards of hygiene in a kitchen:

(a) Are hair and beards tidy and clean?

(b) Do female staff and long-haired male staff wear protective hair covering?

(c) Do all the staff wear neat, well-fitting washable protective workclothes?

(d) Are the dishwashing and vegetable preparation staff supplied with waterproof aprons to keep their workclothes dry?

(e) Do the staff have clean hands? Are all sore places or cuts covered with waterproof dressings?

(f) Are all workclothes clean and in good repair?

(g) Is the state of the staff toilets (WCs) and rest areas good?.

(h) Is the wash-basin in the kitchen clean and is there soap and a nail-brush and a hygienic means of hand-drying?

(i) Are the walls and ceilings clean?

(j) Are the floors and tabletops clean? (some latitude may be allowed at busy meal times)

(k) Is there plenty of light?

(l) Is the ventilation efficient?

(m) Are the storerooms tidy and clean?

(n) Is the garbage collection area clean, fly-screened, tidy?

If the answer to all the above questions is "Yes" it is likely that the kitchen staff work in a hygienic manner.

Regular weekly and monthly inspections of the kitchen premises by the canteen supervisor or manager are a necessary part of maintaining good hygienic standards.

Certain areas of the kitchen and cafeteria should be cleaned daily, weekly, or monthly, and it is good management practice to schedule a regular time and a member of staff for such cleaning. It is better if one person is responsible for the cleaning of a particular piece of equipment or area of the canteen on a regular basis, as this can provide an element of responsibility. It is of course desirable for staff who perform the rather mundane but necessary cleaning tasks to be given recognition and praise when they do their job well.

Choice of food and menu planning

A small canteen catering for ten or twelve workers, all from the same local area and belonging to the same ethnic and religious group will not usually have a choice of menu. The cook will know what foods are liked, how they should be seasoned and cooked to please his customers and will also be ready to make small adjustments to suit individual taste and appetite.

In a canteen catering for 150 or more persons it is unlikely that there will be such a "family" atmosphere, and a simple choice of menu will be appreciated by customers. The cooking equipment available will affect the ability to provide a choice but doing so can help to ensure the most effective use of equipment and available foods. Rather than share out a particular type of food that is in short supply in small portions to everyone, it can be served in normal-sized portions as an alternative to another dish.

Cooking for ethnic groups or tribes with different tastes or taboos or for members of religions with strict dietary laws can often be simplified by providing a choice of menu. This may vary from completely different dishes to a choice of accompaniments or of flavourings. If half the workers are accustomed to eating hot spicy foods and the other half bland dishes, the most satisfactory solution is to prepare a hot, spicy sauce for the first group and a mild bland sauce for the second. Local knowledge is usually adequate to deal with those situations. In a small canteen, word of mouth is enough - "the food you like is at the next counter" - while in larger canteens the foods can be identified by clearly written notices on a blackboard.

Religious dietary laws should be followed as far as practicable. They have usually evolved over generations as a means of preserving public health and their followers survived while others fell ill or died of food-borne infections. Genuine attempts to comply with religious dietary laws are always appreciated. It is also true that, provided the concept of a choice of menu is accepted, it is not difficult to meet the requirements of these laws.

In an area where several ethnic or religious groups coexist it can be helpful to employ someone from each group in the canteen to provide information and advice.

Every establishment will have its own way of dealing with the question of choice, but is is worth remembering that apart from the problem of different religious groups and ethnic groups, who all eat different foods, the provision of a choice is psychologically desirable in the work environment.

When there are food shortages, the choice will be limited. A good cook with imagination and encouragement from management will usually nevertheless be able to introduce some element of choice in the menu, if not every day then at least on some days of the week. In a small canteen it is possible to put a list on a blackboard or large sheet of paper in the cafeteria showing the menu for the next day and to let the workers sign under the dish they want to eat. In this way the canteen supervisor knows how many portions of each dish to prepare and there will be no waste. Where there are many illiterate staff someone can sit beside the list and write down names.

In larger establishments it will be found that it is possible to estimate, with a reasonable degree of accuracy, how many customers will eat each dish.

Meal-times

A main meal is not always necessary in the middle of a shift. In areas where workers walk long distances to work without having eaten, a substantial breakfast in the cafeteria will be appreciated and a light snack at midday may suffice.

Evening and night shift workers may prefer to have a substantial meal at the beginning or end of a shift. Older workers may prefer to eat their main meal with their family before going on evening or night shift and to take light snacks during the shift.

Consultation with the workers' representatives on the canteen management/advisory committee is necessary, and it should be remembered that from time to time changes may be called for. A change in bus or train time-tables can affect workers' meal patterns if they have to travel long distances to work.

Staggered meal service

This depends on the number of meals to be served, but, generally speaking, the modern batch or relay system in which foods are cooked in smaller quantities, at, say, 25 minutes intervals, produces food that is fresher, tastes better, has lost less of its vitamin content and is less likely to become contaminated.

A canteen serving 100 meals during one hour could serve them at the rate of 10 customers per minute, and if all the customers arrived simultaneously the meals could all be served within ten minutes. Thus if all the food is ready five minutes at the most before the first customer arrives, it would not have had to stand for more than 15 minutes.

But if the customers come in at intervals over a period of 50 minutes - the food served to the last-comers would have been kept hot for at least 52 minutes. This is desirable neither from the health and hygiene standpoints nor from that of taste.

If, however, the food is cooked in batches, freshly cooked food could be ready for the late comers 25 minutes after beginning serving. In this case the longest time food would need to be kept hot would be 30 minutes. It is normally possible to estimate the number of customers who will arrive in the first 20 or 25 minutes.

Food purchasing

A large canteen in a town will purchase from established wholesalers. Tenders for the supply of non-perishable goods for a maximum of a year, and perishable goods for shorter periods should be sought. Cost need not be the only criterion. Quality, prompt deliveries and storage facilities are also of importance.

Away from towns and cities the situation is different. In very remote areas the canteen may have to keep at least a month's supply of frozen and non-perishable foods and depend on local merchants for perishable food. Whenever possible, consideration should be given to obtaining the cooperation of local merchants. Even the smallest roadside stall-holder will be able to help if given advice and practical assistance in enlarging his business. When the food hygiene laws are not strictly enforced because of a lack of public health inspectors, the canteen in an industrial establishment must monitor the conditions of work of food handlers and the quality of food handling. This may be done by the medical department. The local butcher may need guidance in cutting the meat according to the canteen's needs, and this can be done by means of a practical demonstration or large charts. Small market gardeners may need help with the purchase of seed or fertiliser but once that hurdle is overcome they will be able to produce the vegetables the canteen needs. It is wise to stipulate (in co-operation with the medical department) what fertilisers are to be used and when. A bakery close at hand, in town or country, can be a great help to a small canteen. They will make bread of the type enjoyed by local people and will be able to adjust supplies at short notice.

Payment for meals

Payment for meals in cash requires the services of a cashier and cash register (except for very small canteens) unless everyone pays a fixed amount equivalent to one coin - a rupee, a riyal, a dollar, fifty pence, a franc, and so on - which is dropped into a cash box with transparent sides as they collect their meal.

Where tickets are used to buy meals they can be sold on pay day, for example in books containing sufficient tickets for the pay period; or tickets can be sold near the factory entrance each morning. Tokens may be sold instead of tickets and collected, like coins, in transparent boxes. For night shifts where there is an automatic vending system, the tokens can be used to obtain a meal from the machine.

Protective clothing

It is important to provide canteen workers with hygienic, well-fitting protective clothing just as is done for workers in, say, a pharmaceutical plant or a steel foundry.

For reasons of hygiene, the clothing should be comfortable, light coloured, easily washable and designed for the work. The chef's traditional outfit of blue and white checked cotton trousers, double-breasted heavy white cotton long-sleeved jacket with a cotton scarf, tall white hat and apron fulfils all these functions very well and this is probably why it remains popular around the world. The jacket gives protection against the heat of the oven and its sleeves protect against splashes of boiling fat. The scarf absorbs perspiration from the chef's face. The apron gives extra protection to the body and can be quickly changed if it gets soiled. The hat is designed to allow free circulation of air above the scalp, some insulation against heat and to prevent hair from falling into the food. Traditionally, the taller the hat, the more senior the chef. Nowadays the hat may be designed to open out flat for ease of laundering or may be made of paper and thus disposable.

Vegetable preparation and pan-washing staff may wear darker clothing such as a boiler suit which must however be easily washable. They should be provided with waterproof aprons of rubber or plastic - at a pinch these can be made out of plastic garbage sacks - for wet work; and heavier cloth aprons for dry work - these can be made out of well-washed jute sacks. Wooden or PVC sole clogs or rubber boots should be provided for these workers if they have to stand on a wet floor.

Dark, usually dark blue, cotton aprons are worn by butchers and those who have to work with raw meat, as they do not show blood stains. They should be washed frequently. Meat workers traditionally wear hats with brims to protect them from over-hanging carcases in cold store rooms and these are popular also with kitchen workers in some places. A washable version with a net crown is available.

All canteen staff who cook, prepare food to be eaten raw (salads, etc.), handle dishes and cutlery after they have been washed, or serve food, should wear light-coloured clothing and protective headwear that contains the hair. Since in some regions of the world colour has great significance, the choice of colour for canteen staff clothing is no simple matter. It should be discussed with the canteen committee and the staff themselves. White, for example, may be an unpopular colour for women in areas where it signifies widowhood.

Women staff should wear protective clothing that is acceptable to their culture, safe and hygienic. Saris and long skirts are unsuitable but tunics and trousers in a traditional style of the area are usually suitable and acceptable.

Cloths for use at the ovens or oven gloves are needed for handling hot pots and pans. Wool is particularly suitable, as it is less flammable than cotton. Workers who have to handle large quantities of frozen goods need insulated gloves (and jackets if they are going in and out of the cold room frequently).

The canteen budget should include the cost of providing hygienic protective clothing, its laundering, repair and replacement. Since kitchens are often warmer than elsewhere, protective clothing should be designed so that it can be worn instead of, not on the top of, normal indoor clothing. Apart from the hygiene and safety aspects, the provision of well-made, appropriate, attractive and comfortable protective clothing has a beneficial effect on staff morale and efficiency in a canteen.

Eating arrangements for canteen staff

Very often the canteen staff can have their own meals in the cafeteria before or after the main meal service. If the supervisory staff in the establishment take their meals at a fixed time or if they sit in a special section of the cafeteria, a canteen manager or supervisor should eat with them. This provides the manager with valuable feedback on customers' reactions to the food and service of the canteen.

Fire prevention

The special fire hazards of kitchens and cafeterias are fires arising from burning fat in ovens and fryers and electrical fires.

In kitchens, carbon dioxide, dry powder or foam extinguishers and small fire blankets should be provided for what is often called "first-aid fire-fighting", i.e. the preliminary action that can be taken by the staff before the fire brigade arrives. Larger fire blankets may also be kept in the kitchen for wrapping around persons whose clothes catch alight (a potential risk in kitchens).

Training in the use of this basic equipment is necessary for kitchen staff, like other members of the workforce. New staff members should be instructed by a senior member of the kitchen staff on the immediate action to be taken in the event of a fire, and they should be included in the next training class. If there is a skeleton night staff in the canteen it is most important that they should know what action to take if a fire occurs. Training prevents panic. A bucket of dry sand can be useful in a kitchen fire but for health reasons it should not be replaced in kitchens by a bucket of soil. By the kitchen telephone the extension telephone number of the establishment's own fire brigade should be posted IN LARGE NUMBERS, or the correct procedure for summoning the local fire brigade. Where the canteen is run jointly by several establishments it is important to ensure that the canteen staff participate in any fire-fighting classes, practices or exercises.

Electrical safety

All electrical equipment in the canteen must conform at least to national electrical regulations and if these do not exist or are considered inadequate, the International Electrotechnical Commission's (IEC) Standards should be used.

No unauthorised person should attempt to repair faulty electrical equipment. Circuits must not be overloaded. It should not be possible to switch on simultaneously two pieces of apparatus which are intended to be used singly.

All control switches should be easily accessible at the front of a piece of equipment. It should never be necessary to reach over a pan of boiling fat to switch off a stove. The main switch should be clearly marked and alongside it should be a pictogram of emergency action to be taken in case of electric shock.

The maintenance engineer or electrical engineer of the establishment, or of one designated establishment within a group, should make regular checks of the safety of the electrical equipment in the canteen. In some countries there may be nationally appointed inspectors who will do this for a small fee and insurance companies may insist that they be used.

First-aid

The first-aid arrangements for the canteen come within the province of the establishment's full or part-time medical officer and his or her staff. Unless the medical department adjoins the kitchen, arrangements will have to be made to have trained first-aiders on the kitchen staff. The medical department may provide this training or a branch of the national Red Cross or Red Crescent Society or another organisation may run classes locally.

There should be a properly stocked first-aid box or cabinet in the canteen (probably in the staff rest-room) and basic instructions in the local language(s) written clearly inside the box or on the door of the cabinet. The names of the trained first-aiders on the canteen staff, covering the full working period, should be posted on the wall beside the box or cabinet, as also should be the telephone number of the medical department or doctor on call. Refresher training should be given regularly, whether or not national regulations so require. The contents of the first-aid box should be checked once a month by a competent person. If the establishment runs or participates in regular emergency training exercises, the first-aiders on the canteen staff should take part.

CONCLUSION

There has been in recent years considerable interest in developing countries, among employers and workers alike, in the provision of industrial (or workers') feeding services particularly in small and medium sized enterprises. Legislation exists in many countries requiring the provision of industrial feeding services but such legislation often applies to enterprises employing 100 or more workers or to workers engaged in dirty, toxic or other harmful processes for whom a messroom is required regardless of the size of the enterprise. This manual, although of a wider interest, has been written primarily with a view to helping employers and workers' organisations in taking more informed decisions in the development, organisation and management of feeding services especially in medium and smaller enterprises. The need for such a manual had been expressed on several occasions including tripartite meetings and workshops.

Given the varieties of situations and problems, this manual inevitably deals with the subject only in broad terms[1] and is intended to help employers and workers to work out solutions for providing an appropriate industrial feeding service. It is also hoped that government officials, especially factory and labour inspectors and public health inspectors whose work brings them into close contact with small and medium-sized enterprises, will find the manual of use.

The manual emphasises the point that an industrial feeding service should be appropriate for the enterprise. This is not always fully appreciated. Sometimes governments, employers or workers think that an industrial feeding service for a small enterprise should be a scaled-down copy of one designed for a large mill or factory employing several hundred workers. Among the considerations and adaptations that may therefore be considered in designing an industrial feeding programme are the following:

(a) Nutrition has to be regarded in the context of the culture, traditions and food readily available in the country concerned. Local conditions may not, in the first stages of setting up a workers' feeding service always permit the achievement of what is <u>desirable</u>. There is therefore a need to ensure that local customs are observed, that changes are introduced gradually and that they are accompanied by public education. Enterprises are also advised and encouraged to seek local advice from their national organisations concerned with labour, social welfare, health and food and nutrition education programmes.

(b) Given the resource and organisational problems particularly facing small- and medium-sized firms, enterprises need to consider various ways of providing feeding services including the following:

- Mess rooms which can be very appropriate for small enterprises and where the workers can bring food from home;

- Mobile food vans to take meals for workers on distant sites or to provide refreshment even within an enterprise;

- Inter-enterprise canteens, "public" restaurants and catering from a central kitchen;

- Low-cost or fair price shops which may be run by an enterprise or set up by the enterprise and run by a trade union or workers' consumer co-operative;
- Catering by local vendors and hawkers who can often be integrated into a small feeding service.

The most appropriate form will depend on the size and type of enterprise as well as local conditions and practices. Because of this and also because the financial burden can be onerous, the provision of workers' feeding services is an area that calls for and lends itself to co-operation between employers and workers. As has been shown in various parts of this manual, there are various ways in which workers and employers can co-operate in the development, organisation and operation of feeding sources.

Going beyond the enterprise, there are at least two points which need to be emphasised. Clearly, suitably trained personnel at all levels are essential to the successful operation of food services and these, unfortunately, are lacking in many developing countries. Therefore, every effort should be made by governments to provide or to encourage management to provide training facilities appropriate in form and content for both workers and managerial staff of canteens, stores, co-operatives and other services.

Indeed as long ago as 1971, the FAO/ILO/WHO Expert Consultation on Workers' Feeding recommended:

> That the administration of laws and regulations concerning workers' feeding be facilitated by a technical agency staffed by appropriately trained profes- sionals at national level. Such an agency should also act as an information centre on the benefits of workers' feeding and provide consultant services on the determination of nutritional requirements and available food supplies, menu planning; the organisation, layout and design of the facility; equipment required, sanitation and hygiene; accounting and control; staffing needs, and training of personnel.

Unfortunately, in many developing countries such agencies do not yet exist, and national authorities may therefore wish to consider their establishment so as to provide advisory or consultancy services for enterprises. International organisations such as the ILO, WHO and FAO with a special interest in nutrition and workers' feeding services can provide interested countries with technical assistance in setting up such agencies.

Finally, the role of government in promoting the provision of feeding facilities and services through the formulation of legislation and guidelines should be noted. Legislation can and should provide guidelines on when and where and in what way employers need to provide feeding programmes for workers. The labour and factory inspectorate could play an important role both in supervision and in providing advisory services.[2] In addition since governments are among the major employers, they could and should articulate appropriate standards and models to follow and set an example of the various ways in which workers and employers can collaborate in the development, organisation and management of canteens and food services. In other words, although legislation, suited of course to local conditions, is important, it is not an end in itself; nor is it sufficient by itself. It should be complemented both by effective enforcement and by measures that provide motivation and assistance to both employers and workers in the development and operation of industrial feeding services.

Notes

1 For those interested in more information, a reading list is included as an annex. The list includes publications which are easily obtainable. Readers are also recommended to enquire at their local ILO, FAO and WHO offices or local sales agent for details of recent publications on the subject and of any books published specifically for their country or region.

2 A Check List drawn up to enable government inspectors, management and worker's representatives to check on the more obvious health, hygiene and safety aspects of a kitchen and cafeteria/dining hall is included as Annex. It must be emphasised that this check list has been drawn up for small to medium-sized industries in developing countries only. It is not intended for inspecting canteens in large factories or mills; nor is it intended for use in industrialised countries for which there already exist more detailed and comprehensive check lists.

ANNEXES

Annex I

CHECK LIST

Clean, well-designed and properly maintained premises are essential factors in the hygienic handling (preparation, cooking and serving) of food and in ensuring safe working conditions for food handlers (all kitchen workers from supervisor to floor cleaner should be regarded as food handlers). Kitchens and canteens, whether small or large, are parts of a factory, and the health, safety and welfare of food handlers are therefore the responsibility of the employer just as though they were working on the shop-floor or assembly line. These and other concerns will normally be considered at the planning and building stage. It is possible however, and not usually difficult or expensive, to improve existing premises. The Check List draws attention to such points. It can also serve as a "pocket reminder" for management, workers' representatives and government inspectors when they visit industrial feeding services and when they consider improvements.

The Checklist has been designed so that "Yes" answers are correct.

I. SAFETY OF COOKING EQUIPMENT

Mechanical and electrical equipment

This includes equipment such as mincers, slicers, chippers, vegetable peelers, mixers, shredders, crumb-makers, waste disposal machines, garbage compactors, water boilers, electric ovens, platewarmers, trolleys with electrical connections, ovens, boiling rings, dishwashers, refrigerators, deep-freezers, fans (ceiling, circulating and extractor), air-conditioners, room heaters, light fittings, trolleys and wheeled tables.

- Are all the dangerous parts of the machines properly guarded?

- Is electrical equipment properly insulated and, if necessary, earthed?

- Are all the machines and equipment in working order?

- Are all the machines and equipment properly maintained and regularly serviced?

- Has electrical installation been carried out properly and in accordance with national standards?

- Are the main electric switches in safe but accessible places?

- Is there regular servicing of equipment?

- Is there fire-extinguishing equipment suitable for use on electrical fires?

- Is there a prominent notice in the local language with clear diagrams and pictures showing the emergency action to take in case of electric shock?

- Are all fan blades properly guarded?

- Are the controls on tilting kettles or pans in good order?

Steam-heated equipment

This includes water-boilers, steam ovens, steamers, steam-jacketed boiling pans.

In many countries it is a legal requirement that steam-heated equipment shall be regularly inspected by an insurance company's inspector or a government boiler inspector and that records are kept.

Even where there is no legal requirement it is advisable that steam equipment should be regularly inspected by a competent person and records of such inspections kept on the premises.

- Is there adequate lagging (insulation) on the equipment?

- Is the lagging (insulation) in good condition?

- Are there sufficient safety valves?

- Are there automatic cut-off devices on oven and steamer doors?

- Is steam equipment regularly maintained and serviced?

- Is steam equipment regularly inspected by an insurance company inspector or a government boiler inspector?

- Are government requirements as to inspections, records, etc., fulfilled?

Gas equipment

This includes cookers, boiling tables, hot-plates, rings, griddles, grills, salamanders, deep-fryers, Bratt pans, ovens, boilers, tilting pans, refrigerators, deep-freezers, and room heaters run on gas, either bottled or town (piped) gas. It may be a legal requirement to employ only approved gas fitters for installation, servicing and repairs.

- Is there sufficient lagging or insulation to prevent workers getting burns from accidentally touching equipment?

- Is all the lagging and insulation in good order?

- Is equipment properly maintained and serviced?

- Are there automatic cut-off devices if a pilot light goes out?

- Are containers of "bottled" gas stored in a safe place outside the kitchen?

- Have workers been trained in the safe way to handle gas containers?

- Are there close-fitting, air-excluding covers for fryers to be used in case of fire?

See also section on VENTILATION.

Kerosene cookers
See under VENTILATION

Solid fuel cookers
See under VENTILATION

II. LAYOUT OF PREMISES

Are there at least two doors 0.75m (30 in) wide, each on a different wall, opening outward and leading to the outside, to serve as emergency exits in the kitchen building?

- Is the space in front of and behind the emergency exit doors always kept clear to permit quick evacuation of persons in case of emergency?
- Is there adequate space around the stoves for two workers to pass?
- Is there sufficient space between equipment for two persons to pass even when refrigerator, oven or cupboard doors are open?
- Is the equipment the right size for the workers who use it?
- Is visibility in the kitchen good with no blind corners?
- Are the toilets separate from but not more than 30-40 m (100-140 ft) from the kitchen?[1]
- Are there adequate toilet facilities?
- Are they clean and in good repair?
- Are there facilities for handwashing outside the toilet(s)?
- Is soap provided for handwashing?
- In cold climates is there an approved means of hand-drying for food handlers? (Not essential in warm climates)
- Is there at least one hand-wash basin in the kitchen?
- Is it reserved exclusively for hand-washing?
- Is there a hand-wash basin available for customers?[2]

Special problems met in rural areas

Mobile and "temporary" canteens feeding workers on temporary work in rural areas, such as dam building and construction work or in plantations, face different problems from those in industrial areas. But even in these situations it is possible to produce good meals and achieve reasonable standards of hygiene and safety. The same general guidelines apply. If solid fuel or bottled gas are the available cooking fuels, it is still important to store the fuel away from where food is prepared, cooked or served and to ensure that there is good ventilation, often achieved in a temporary canteen by cooking out of doors or under an open-walled shelter.

The provision of water for hand-washing for food handlers and for customers who prefer to eat with their hands is as important on a building site as in an engineering works in a city. Water may have to be brought in metal containers or cisterns from a stand-pipe or well or a tube-well may have to be sunk by the contractor.

Cess-pits may have to replace water-borne sewage systems. Chemical closets, privies and latrines may have to replace water-closet toilets.

Points to check include:

Where there is no running water:

Hand-washing

- Is there a covered metal container with a tap, e.g. a barrel, drum or water-butt, raised above the ground on stones, on a table or on the tailgate of a lorry outside the kitchen or mobile canteen?

- Has the tap a long wooden or metal extension which can be operated by the user's elbow?

- If latrines are provided do they meet best local standards?

Remember that where there is not a water-borne sewage system any chemical closets, latrines or privies should be at least 30 m (100 ft) from a kitchen or any place where food is stored or served.

- If there is a cess-pit is it in working order?

- Are cess-pits and latrines sited on lower ground than the kitchen and/or canteen?

- Is there a fence round any masonry wells to keep animals and young children out?

- Is there potable water (safe drinking water) available either from a tap, a water cooler or from a covered container with a tap?

- If the water is not taken from the mains supply is it sent at regular intervals to an approved laboratory for testing?

Floors, walls and ceilings

- Is the floor of impervious, non-slip material?

- Has the floor been laid with a slight fall (slope) toward any drains and gullies?

- Are all open gullies and drains covered with gratings in good condition?

- Are all drains and gullies leading to the outside protected with small mesh wire-netting to keep animals out?

- Are all grease traps properly covered?

- Are gratings and covers on gullies, drains and grease traps easy to remove for cleaning?

- Are the floors, walls and ceilings clean?

- Are the walls and floors free from cracks?

- Can walls and ceilings be washed easily?

- Are there long-handled hoses, brushes, and squeegees available for cleaning the ceiling and upper walls?

- Are any fans (circulating, extractor or ceiling) clean and in working order?

- Are extractor fans properly ducted (piped) to the outside?

- Are hoods over stoves and other cooking equipment clean?

- Are there fly-screen doors of wire or nylon mesh in good repair to keep out flying insects?

Sinks

- Are the sinks in reasonable condition and free from cracks?

- Are they clean?

- Are there enough sinks?

- Is the sink(s) used for vegetable or meat or fish cleaning distinct and separate from those used for dishwashing?

- Are there duckboards in front of sinks so that the workers do not have to stand in water as they work?

Windows

- Are they clean?

- Are they fly-screened?

- Can they be opened easily and safely?

- If curtains are fitted are they made of flameproofed materials?

Garbage disposal

- Have all garbage containers got well-fitting covers?

- Has the storage area for garbage containers outside the kitchen got a concrete, tiled or brick floor?

- Are the containers raised above ground?

- Is the storage area fly-screened?

- Is there a fence or wall to keep animals and vermin out of the storage area?

- Is there easy access to the area for lorries or carts coming to collect garbage and swill (for animal feeding)?

- Is the garbage separated into clearly marked containers[3] for example:

  one container for food (for animal swill);
  another for clean paper;
  another for glass; and
  one for other items?

III. VENTILATION

Good ventilation plays an important part in maintaining efficiency and hygiene in kitchens. Hot air and fumes should be removed from the atmosphere and replaced by cool clean fresh air. This is most commonly done by using extraction (exhaust) fans. If electric power is not available or inadequate, full use should be made of natural ventilation and for this there must be plenty of smaller windows (with outside shutters) so arranged to catch the prevailing winds (which may change with the season).

Note that where extractor fans are provided, if there is a strong smell of food cooking in the cafeteria/dining hall but not in the kitchen, immediate attention should be given to the ventilation system. Normally air is pulled from the cafeteria into the kitchen.

Table or wall mounted circulating fans are necessary in warm climates unless desert-type fan-coolers or air-conditioning are installed. Modern ceiling (punkah) fans also improve working conditions and efficiency in kitchens.

Special care must be taken to provide good and efficient ventilation when solid fuel, kerosene or gas are used for cooking.

Kerosene

Is ventilation adequate without risk of excessive draughts which could cause a flare up?

Is the kerosene stove standing on a tray with a deep edge or is it standing inside a low-walled brick or concrete enclosure wall, so that if the stove overturns any spilt kerosene will be confined to a relatively small area?

Is the kerosene stored in a safe place outside the kitchen?

Solid Fuel (charcoal, wood or coal)

Where solid fuel, especially charcoal, is used there must be extra ventilation in the kitchen. Points to be checked where solid fuel is used include:

- Are the flues (chimneys) of the stoves cleaned (swept) regularly?
- Are the flues in good repair?
- Is the fuel store outside the kitchen?
- Is there a rule that charcoal must never be used for cooking indoors but only on a well-ventilated verandah or in the open air?

Gas appliances

- Is there a flue (chimney) in good repair for each item of gas equipment?
- Is there a flue in good repair for a "mixed" cooker or boiling table using both electricity and gas?

IV. FALLS AND TRIPPING ACCIDENTS

To reduce the number of falls and tripping accidents in kitchens it is necessary to take certain relatively simple and inexpensive precautions and to make regular checks on the following questions:

- Is the floor clean and free from grease and other spills?
- Are a sweeping brush or broom, a dust pan and a mop and bucket, kept handy in the kitchen so that spills can be swept or mopped up quickly?
- Are the passages between equipment and doors (especially Emergency Exits) kept free from obstructions such as boxes and sacks?
- Is there sufficient space between equipment for two people to be able to pass easily even when oven, refrigerator and cupboard doors are open?
- Are all cupboards and shelves at a height which can be easily reached by any member of the kitchen staff?
- If shelves and cupboards are out of reach for any workers in the kitchen is a strong step-ladder provided?
- Is it always possible for a worker pushing a trolley or carrying a tray to see ahead?
- If not, has a mirror been put up so that workers carrying a tray or pushing a trolley can avoid bumping into another person?

V. PERSONAL HYGIENE FACILITIES FOR FOOD HANDLERS

- Are food handlers provided with light-coloured, easily-washable aprons, protective coats, overalls, trousers, caps, etc.?
- Are these garments clean and in good repair?
- Is there a small hand-wash basin with soap in the kitchen?

- In hot climates or where workers' homes are lacking facilities, is there a washroom or shower-room for the use of kitchen workers?

VI. CAFETERIAS, DINING HALLS AND MESS ROOMS

It is an advantage to have flexible seating arrangements (perhaps with stacking chairs and tables) so that the room may be used outside of mealtimes for activities such as union meetings, health education classes, table tennis, training sessions, yoga classes, discussions and social groups to which workers or their families belong.

- Is there extra seating accommodation, perhaps 5 to 10 per cent extra, so that customers can easily sit with friends or colleagues at mealtimes?

- Are the tables and chairs easily movable?

- Is the ventilation good?

- Is there adequate equipment provided for heating food brought from home or for boiling water in the mess room?

- Is the lighting, especially over the serving counter good?

- Are the floor and tables clean?

- Are the walls, windows and ceilings clean?

Vending machines

- Are they in good working order and regularly serviced?

- Are they clean?

Trolley services

- Is the equipment, trolley, water boiler, multipot, etc, clean and in good working order?

- Are wheels on trolleys and tables, etc., strong enough for the purpose?

- Are there effective brakes on trolley and table wheels?

- Do trolleys and other wheeled equipment move freely when the brakes have been released?

Insulated food and beverage containers

- These may be used for transporting meals and beverages to satellite canteens and isolated work sites and in central kitchens.

- Are they clean and in good condition?

- Do the lids fasten properly and is the insulation intact?

- Are the vehicles, i.e. vans, flat-bed lorries, pick-ups, handcarts, rickshaws, used to transport containers clean?

- Is the vehicle driver provided with a clean cloth or gloves for handling the containers?

VII. FIRST-AID

- Is there adequate first-aid equipment (including an instruction chart) and does it meet national regulations?[4]

- Do any of the canteen staff hold up-to-date first-aid certificates?

- Do holders of first-aid certificates take refresher courses?

- Is there a prominent notice in languages understood by the workers with clear diagrams and pictures showing the emergency action to take in case of electric shock?

- Is there a book in which first-aid treatments given are noted?

- Is there a book in which any accidents occurring in the kitchen are noted down?

VIII. FIRES AND EMERGENCIES

- Are there adequate hand-held fire extinguishers?

- Is there a fire extinguisher(s) suitable for electrical fires?

- Is there a prominent notice in languages understood by the workers with clear diagrams and pictures showing the emergency action to take in case of electric shock?

- Is there a bucket of sand and a fire-blanket?

- Does the fire-fighting equipment meet national regulations?

- Have all canteen staff and workers attended at least one fire-drill or training session on the use of extinguishers, etc.?

- Where there is a mains (piped) supply of water, is there a hosereel with the correct coupling/fitting for the hydrant?

- Are there wall-posters with clear instructions - in languages understood by the workers and in pictures - on what to do if there is a fire?

GENERAL IMPRESSIONS

It is important to be able to judge between the temporary apparent confusion of the final few minutes before a meal is served in a well-run canteen and the permanent untidiness and lack of cleanliness found in poorly-run canteens. It is very difficult for a busy kitchen to be immaculate all day every day but it should always be tidy and the staff always neat and clean in appearance. At least once in every shift (which means at least three times in 24 hours in a round-the-clock operation) there will be a period when the well-run kitchen, whether it be large or small, modern or old, in a town or in the country, will be thoroughly cleaned. Such thorough cleaning of premises combined with hygienic food handling by the staff of an industrial feeding service plays a great role in the nutrition and health of workers.

Notes

[1] Note that where there is not a waterborne sewage system the chemical closets, latrines or privies should be at least 30 m (100 ft) from a kitchen or any place where food is stored or served, and if the site is sloping the kitchens and food stores should be higher up the hill or slope than any toilet facilities.

[2] Important where customers eat with their hands.

[3] This is done:
(a) to prevent broken glass or pottery injuring a garbage collector, and
(b) to enable garbage to be sold or used in official recycling schemes.

[4] If there are no national regulations on first-aid equipment, the national Red Cross or Red Crescent Society can provide a basic list and may also run basic first-aid courses in local languages.

Annex II

COOK/FREEZE AND COOK/CHILL METHODS

The up-grading or introduction of an industrial feeding service for the workers in small to medium-sized establishments is usually on a small scale and involves traditional methods of cooking. New developments are, however, taking place rapidly in many industrialised countries and some mention is therefore made of them in this annex.

The aims of the cooking methods known as "cook/freeze" and "cook/chill" are to utilise more effectively the kitchen staff by using the skilled cooks for the work they do best and using semi-skilled workers for less skilled operations such as salad-making, serving food, vegetable cooking, etc. Equipment can also be used to best advantage and in cook/freeze it is possible to take advantages of the availability of seasonal abundances and consequent lower food prices. Both methods permit the centralisation of the major part of the cooking for a number of canteens, oil rigs, hospitals, airlines, hotels, etc.

In areas where labour is costly and there are few skilled workers these methods have many advantages and are said to be well suited to feeding shift workers, provided there are suitably trained canteen staff available to provide the finishing touches and modifications to make the food attractive and enjoyable. It is possible to issue a menu either the previous day or as the workers come on shift offering a variety of dishes from the freezer from which each worker can choose and place an order. At the meal break the meal ordered can be ready for each worker - freshly reheated (regenerated) from the freezer.

In isolated areas where the workers have no possibility of eating anywhere but in a company cafeteria or messroom, the use of cook/freeze or cook/chill for all the canteen catering can give rise to complaints of monotony. It is better in such cases to make a certain amount of use of traditional cooking methods to supplement the cook/freeze or cook/chill methods. A grill or barbecue where customers can see "their" meals, be it kebabs, grilled fish or bacon and egg being cooked for "them" alone makes the customer feel an individual, not just one of 100 people who happen to work on the same site and have to eat together.

Cook/freeze operations involve a kitchen or part of one being converted into a small food "factory", run on a single or double shift according to the numbers catered for.

Foods suitable for deep-freezing are cooked ahead of requirements, packed into suitable sized containers, either re-usable aluminium or stainless steel, disposable aluminium foil containers, plastic bags or disposable paper/plastic laminated containers.

The foods are then pre-cooled in a cool room with circulating fans, then put into a blast-freezer which freezes them in approximately 2 hours to at least - 20°C (4°F). The containers, clearly marked to indicate their contents and the day of cooking, are then put into a deep freezer for storage at - 20°C (4°F) until required.

The frozen foods are regenerated for serving either in - (a) a forced convection oven, (b) a microwave oven, (c) steam oven, usually high pressure, or (d) for plastic bag packaging only, in boiling pans. Since not all foods can be frozen it is necessary to have a finishing kitchen(s) alongside the cafeteria(s) where salads, fresh fruits, certain fried foods such as chips and some sweets are prepared.

A modification of the system is to cook all foods for the main meals freshly but to cook extra portions of those dishes which freeze well and freeze these in single portions as a reserve for days when there is an unexpected number of customers (as the single portions can be quickly regenerated in a microwave or steam oven). Where there is a small night shift, the frozen meals can be used for a self-service meal if there is a small freezer, preferably with a glass door so that the contents are visible, and a microwave oven.

Careful record-keeping and planning of cook/freeze operations are essential and hygiene standards must be high.

<u>Cook/chill</u> operations, though still involving a factory type of production, are based on keeping food no more than five days (including the day of preparation and the day of eating) in chilled storage at a temperature between 0°C (32°F) and 3°C (37°F).

The food is cooked, placed in shallow metal dishes or disposable aluminium containers, pre-cooled and blast chilled to 3°C (37°F) in approximately 1 1/2 hours. To ensure the food is chilled completely, temperature probes take the temperature in the core of the product - it should read between 2° and 3°C (35-37°F).

Once chilled the food is transferred to a refrigerated holding room or cupboard maintained at 3°C (37°F). All food is marked with the date of production and the date of proposed consumption. Outdated food (i.e. more than 5 days since cooking) has to be discarded.

The food is regenerated when required as in cook/freeze systems. If the food has to travel, say to an outlying cafeteria, it must be kept below 3°C (37°F) during the journey which involves the use of very well-insulated containers or - more usually - refrigerated vans. If the food has to be plated before regeneration the room in which it is plated must be kept at a temperature below 10°C (50°F) so the staff in the plating room must wear insulated protective clothing.

Cook/chill uses one third the amount of electricity used for cook/freeze and enables a large catering establishment to make best use of its skilled cooks while semi-skilled staff serve food, make salads and sandwiches and prepare fruits.

Power supply

Both these methods require a sure and certain electricity supply with back-up generators in case of power outages. If town gas is used as the refrigerator fuel (electricity will still be needed for fans, etc.) there must be a bottled gas back-up or a diesel generator.

Annex III

BIBLIOGRAPHY

Aylward, F.: The feeding of workers in developing countries, Food and Nutrition Paper No. 6. (FAO, Rome, 1976).

Aylward, F. and Jul, M.: Protein and nutrition policy in low-income countries, FAO Energy and Protein Requirements Committee, Rome (Charles Knight, London, 1975).

Brooks West, B., Wood, L., Harger, V., and Shugart, G.S.: Food service in institutions (John Wiley, New York) Use most recent edition.

Busztyn, P.G.: "A diet survey in Zimbabwe", in Human nutrition: Applied nutrition, 39A, 1985.

Department of Employment and Industrial Relations, Australia: Food at work: Use and care of food service equipment, Occupational Safety and Health Working Environment Series 27 (Government Publishing Service, Canberra).

---: Codes of practice: 310 Food service, Hygiene; 311 Operations; 305 Personal facilities (Working Environment Branch, Canberra, 1983).

Department of Health and Social Security, United Kingdom: Clean catering - a handbook on premises, equipment and practices for the promotion of hygiene in food establishments (HMSO, London, 1985).

---: Hygiene in microwave cooking, Food Hygiene Codes of Practice, No. 9 (HMSO, London, 1985).

Durnin, J.V.G.A.: Synthesis: Inter-relationships between protein requirements and physical activity, in "Alimentation et Travail", Deuxième Symposium International, May 1974, Vittel, France.

FAO/ILO/WHO: Report of the Expert Consultation on Workers' Feeding, 10-15 May 1971, Rome, in Nutrition Miscellaneous Meetings Series, No. 2 (Rome, 1971).

FAO: Food composition tables for use in Africa (Rome, 1968).

---: Food composition tables for use in East Asia (Rome, 1972).

---: Food composition tables for the Near East, Food and Nutrition Paper No. 26 (Rome, 1982).

Food Research Institute, Ghana: Food Composition Tables (Accra, Ghana).

International Electrotechnical Commission (IEC): Catalogue of Publications (Geneva). Use most recent edition.

ILO: Welfare Facilities Recommendation, 1956 (No. 102).

---: Occupational Health Services Convention, 1985 (No. 161).

---: Model Code of Safety Regulations for industrial establishments for the guidance of governments and industry (Geneva, 1962), pp. 417-422.

---: Safety and health in building and civil engineering work (Geneva, 1982) Ch. 37, Hygiene and Welfare, pp. 335-340, and Ch. 39.3 Catering, p. 350.

---: Encyclopaedia of occupational health and safety (Geneva, 1983), Articles on - "Beverages at work" pp. 269-270, "Catering, industrial (operation)" pp. 427-432, "Catering, industrial (organisation)", pp. 432-433, "Food-borne infections & intoxications", pp. 898-900, Sanitary facilities, pp. 1998-2002, Nutrition and Diet, pp. 1482-1485, Emergency Exits, pp. 873-877.

---: Tasks to jobs - Developing a modular system of training for hotel occupations, Hotel & Tourism Series, No. 3 (Geneva, 1979).

---: Hotel and Tourism Occupations - Directory of Training Facilities (Geneva, 1976). In the process of being revised and updated.

ILO/MATCOM: Basic rules for the display of goods, Element No. 13-02 (Geneva, 1979).

---: The display of goods in self-service shops, Element No. 13-03 (Geneva, 1979).

---: Buildings and equipment for small shops, Element No. 13-01 (Geneva, 1979).

---: Planning and controlling the business, Element No. 12-03 (Geneva, 1979).

---: Basic economics of a consumer co-operative, Element No. 12-02 (Geneva, 1979).

---: Stock-taking, No. 18-03 (Geneva, 1979).

---: Weighing and pre-packing, Element No. 14-04 (Geneva, 1980).

---: Handling of cash, Element No. 17-01 (Geneva, 1980).

---: Cash control without a cash register, Element No. 17-02 (Geneva, 1980).

---: Cash control using a cash register, Element No. 17-03 (Geneva, 1980).

---: Receipt of goods, Element No. 15-02 (Geneva, 1980).

---: Price-marking, Element No. 15-05 (Geneva, 1981).

---: Stock control records, Element No. 18-02 (Geneva, 1981).

---: Financial management - Material for management training in agricultural co-operatives (Geneva, 1983).

INCAP-ICNND: Food composition tables for use in Latin America, (National Institute of Health, Baltimore, Maryland, 1961).

International Organization for Standardization (ISO): Catalogue of Publications (Geneva). Use most recent edition.

Joy, L. and Payne, P.: Food and Nutrition Planning Report, Nutrition Consultants Report, Series No. 35 (FAO, Rome, 1975).

Kinton, R., and Ceserani, V.: The theory of catering, fifth edition (Edward Arnold, London, 1985).

Kotschevar, L.H.: The feeding of industrial workers in the Near East, mimeographed report of joint WHO/FAO survey in Jordan, Iran and Pakistan (FAO, Rome, 1964).

Kotschevar, L.H. and Terrel, M.E.: Food service layout and equipment planning (John Wiley, New York, 1961).

Lawson, F.: Principles of catering design (Architectural Press, London, 1978).

Marks, J.M.L., Reiner, M. and Ahmad, K.F.: Nutrition Manual for East Pakistan (WHO/FAO/UNICEF, Dhaka, 1965).

Maturu, N. Rao: "Nutrition and labour productivity", in International Labour Review (ILO, Geneva) Vol.118, No. 1, Jan/Feb. 1979.

Micuta, W.: Modern stoves for all (Intermediate Technology Publications, London and Bellerive Foundation Geneva, 1985).

National Standards Institute: Standards for kitchen and dining room equipment (ANSI, New York, 1981).

NINI (National Institute of Nutrition, India): Nutritive values of Indian foods (Hyderabad, India, 1981).

---: Recommended dietary intakes for Indians (Hyderabad, 1984).

---: A manual of nutrition (Hyderabad, 1984).

---: Survey on 8,000 urban households in India (National Nutrition Monitoring Bureau, Hyderabad, 1985).

---: Menus for low-cost balanced diets and school lunch programmes: Suitable for South India and recipes suitable for North India (Hyderabad, 1984).

Oomen, H.A.P.C. and Grubben, G.J.H.: Tropical leaf vegetables in human nutrition, Department of Agricultural Research, Netherlands, Communication No. 69 (Amsterdam, 1977).

Pasricha, S. and Rebello, L.M.: Some common Indian recipes and their nutritional values (NINI, Hyderabad 1984).

Passmore, R., Eastwood, M.A. et al.: Davidson and Passmore, Human Nutrition and Dietetics, Eighth edition (Churchill Livingstone, Edinburgh, 1986).

Paul, A.P.: Nutrition and productivity in India (A preliminary study) (ILO, New Delhi, 1984).

Platt, B.S.: Table of representative values of foods commonly used in tropical countries, Medical Research Council Special Report Series, No. 32 (HMSO, London, 1962).

Queen's College, Glasgow et al.: A question of balance: ergonomics in staff restaurants (Sutcliffe Catering Group Ltd., London, 1986).

Sell's Publications, Ltd., United Kingdom: A Buyers Guide to Equipment and Services to the Hotel and Catering Industry (Epsom, United Kingdom). Published annually.

Sen, R.N.: "Certain ergonomic principles in the design of factories in hot climates", in Occupational safety, health and working conditions and the transfer of technology (ILO, Geneva, 1982).

Tomkins, A.: "Nutrition and health programmes - do they work?", in Nutrition News, Vol. 62, 1985 (NINI, Hyderabad).

Truswell, A.S.: "ABC of nutrition", in British Medical Journal, 1985 (a series of articles).

Wharton, P.A. et al.: "Sorrento Asian food tables", in Human Nutrition: Applied Nutrition 37A, pp. 378-402 (London, 1983).

WHO: "Food Composition Table for the Western Pacific Region", in The health aspects of food and nutrition (Manila, 1979).

---: Mass Catering by R.H.G. Charles. European Series No. 15. Copenhagen, 1983.

Journals

CATERER AND HOTELKEEPER. Weekly. Published by Business Press International Quadrant House, The Quadrant, Sutton, Surrey, SM2 5AS, UK. Includes INDUSTRIAL CATERER supplement every two months.

FOOD AND NUTRITION. Twice-yearly. Published by FAO, Rome. Obtainable from FAO's local agents or Regional Office and can be paid for in local currency.

HOSPITALITY. Monthly. Published by Hotel, Catering and Institutional Management Association, 191 Trinity Road, London SW17 7HN, UK.

ICA COMMUNICATOR. Quarterly. Published by Industrial Catering Association, 1, Victoria Parade, By 331 Sandycombe Road, Richmond, Surrey, UK.

THE INDIAN JOURNAL OF NUTRITION AND DIETETICS. Monthly. Published by The Indian Journal of Nutrition and Dietetics, Sri Avinashilingam Home Science College for Women, Coimbatore - 641 043, S. India.

ZIMBABWE HOTEL AND CATERING GAZETTE, P.O. Box 8045, Causeway, Harare, Zimbabwe.

www.ingramcontent.com/pod-product-compliance
Ingram Content Group UK Ltd.
Pitfield, Milton Keynes, MK11 3LW, UK
UKHW061817190726
13853UKWH00006B/2199

9 789221 066378